Engineering Chemistry

Other Titles of Central West Publishing

Biopolymers Based Advanced Materials
ISBN: 978-0-6482205-4-1 (e-book)
ISBN: 978-0-6482205-5-8 (hardcover)

Functional Polymer Blends and Nanocomposites
ISBN: 978-0-6482205-6-5 (e-book)
ISBN: 978-0-6482205-7-2 (hardcover)

Polymers in Oil and Gas Industry
ISBN: 978-0-6482205-0-3 (e-book)
ISBN: 978-0-6482205-1-0 (softcover)

Functional Nanomaterials and Nanotechnologies: Applications for Energy & Environment
ISBN: 978-0-6482205-2-7 (e-book)
ISBN: 978-0-6482205-3-4 (softcover)

Technology Management in Business
ISBN: 978-1-925823-02-8 (softcover)

Advances in Polymer Technology: Material Development, Properties and Performance Evaluation
ISBN: 978-1-925823-00-4 (e-book)
ISBN: 978-1-925823-01-1 (hardcover)

Polymer Nanomaterials for Specialty Applications
ISBN: 978-1-925823-03-5 (e-book)
ISBN: 978-1-925823-04-2 (hardcover)

Advanced Materials
ISBN: 978-1-925823-05-9 (e-book)
ISBN: 978-1-925823-06-6 (hardcover)

Engineering Chemistry

K. Dinakaran
G. Sasikumar

CWP

Central West Publishing

NATIONAL LIBRARY OF AUSTRALIA

A catalogue record for this book is available from the National Library of Australia

ISBN (print): 978-1-925823-10-3

Contents

Preface

We are delighted to present the Engineering Chemistry book to the readers. Especially, the book has been developed as per the content of the course Engineering Chemistry, which is common to all engineering branches in major national and international universities. We hope that the students of all academic backgrounds will find this book logical and easy to understand. The book has been specifically written for bachelor of engineering (BE) and bachelor of technology (BTech) students to acquire knowledge and understanding about the basics and applications of chemistry in engineering and technology.

We have tried our best to make the topics simple and easy to follow. The book includes diagrams, abundant data and solved problems to reinforce the concepts. A question bank is also included at the end of the book for the benefit of the students.

Constructive suggestions for improving the book are welcome from readers.

K. Dinakaran
G. Sasikumar

UNIT I

WATER AND ITS TREATMENT

I.1 Introduction

The hardness of water is referred to the amount of dissolved salts of calcium (Ca) and magnesium (Mg) in water. If the amount of dissolved calcium and magnesium salts is higher, the water is said to be hard water and water with lower content of Ca and Mg salts is referred to as soft water. The calcium and magnesium are present in water in the form of their carbonates, sulphates, chlorides and hydroxides and their cations have the charge of 2+ (Ca^{2+} and Mg^{2+}). Iron, aluminium and manganese can also be present in higher concentrations in certain cases and these cations also contribute to hardness of water. Though hard water has limited health benefits, it causes severe problems in industrial and domestic usage.

I.2 Disadvantages of Hard Water

I.2.1 In Domestic Usage

For washing and bathing, hard water creates difficulties as it produces thick **scale** in kettles and water heaters. Importantly, hard water does not produce **lather** with soap solution because the surfactant molecules in soap react with the calcium/magnesium in water and produce "soap scum", as shown in the following equation

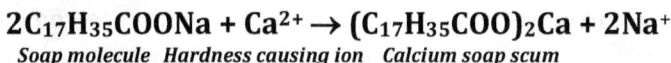

$$2C_{17}H_{35}COONa + Ca^{2+} \rightarrow (C_{17}H_{35}COO)_2Ca + 2Na^+$$
Soap molecule   Hardness causing ion   Calcium soap scum

It also creates sticky precipitates that deposit on bathtubs, body, clothes etc. For cooking, hard water creates similar difficulties by producing scum at the bottom of the vessels. Hard water causes unpleasant taste to food, tea and coffee. Drinking of hard water for prolonged duration may adversely affect the digestive system. In addition, the possibility of deposition of calcium oxalate crystals in the urinary tract cannot be neglected.

I.2.2 In Industrial Usage

1) Boilers (used to generate steam): the use of hard water creates many problems like (i) scale formation, (ii) corrosion, (iii) priming and foaming and (iv) caustic embrittlement. The dissolved salts elevate boiling point and significant amount of precious fuel is wasted during steam generation process.

2) In textile and dyeing industries, hardness causing salts interfere with the effective dyeing and printing of the fabrics. The present of iron salts also results in undesirable shade on fabrics.

3) Paper industry: smooth and glossy finish of the paper is affected by the presence of Ca^{2+} and Mg^{2+}.

4) Sugar industry: the dissolved salts in water co-precipitate with sugar during crystallization and make sugar refining more difficult.

5) Hard water is not suitable for preparing drug solutions in pharmaceutical industry.

6) The hydration of cement and final hardening of cement are affected by use of hard water in concrete making.

I.3 Types of Hardness

I.3.1 Temporary Hardness (Carbonate Hardness)

Temporary hardness can be removed/reduced by boiling the water and by the addition of lime (calcium hydroxide). Temporary hardness is caused by the presence of bicarbonate salts such as calcium bicarbonate and magnesium bicarbonate. Upon boiling the water, calcium bicarbonate and magnesium bicarbonate are converted to their carbonate and precipitate out of water, since calcium carbonate and magnesium carbonate are insoluble in water (equation (a)). Similarly, the addition of lime also produces carbonate and hydroxide salts, which precipitate out of water, as shown in equation (b).

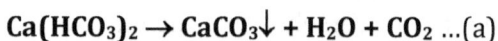

$$Ca(HCO_3)_2 \rightarrow CaCO_3\downarrow + H_2O + CO_2 \text{ ...(a)}$$

$$Mg(HCO_3)_2 + 2Ca(OH)_2 \rightarrow Mg(OH)_2\downarrow + 2CaCO_3\downarrow + 2H_2O \text{ ...(b)}$$

I.3.2 Permanent Hardness

Permanent hardness is the hardness that cannot be re-moved/reduced by boiling. Permanent hardness is caused by the presence of sulphate, chloride and nitrate salts of calcium /magnesium in water. The sulphate, nitrates and chloride salts do not precipitate out upon boiling. Permanent hardness can be re-moved using water softening methods such as ion exchange columns, reverse osmosis, coagulation, etc.

Overall, total hardness and water softening are defined as follows:
Total hardness = permanent hardness + temporary hardness
Water softening: reducing/removal of the salt content of water is known as water softening.

I.4 Units

The total water hardness is the sum of the molar concentrations of Ca^{2+} and Mg^{2+}, in mol/L or mmol/L (millimols/litre) units or in parts per million (ppm). Water hardness determines the total concentrations of calcium and magnesium. However, as mentioned earlier, iron, manganese and aluminum salts can also be present in higher concentrations.

The temporary and permanent hardness are also expressed in $CaCO_3$ scale. The choice of $CaCO_3$ is due to the fact that its molecular weight is 100 and equivalent weight is 50 and it is the most insoluble salt in water. Equivalent of $CaCO_3$ hardness is principally expressed in ppm unit. Other units such as French degree of hardness, English degree of hardness or Clark, USA degree of hardness and German degree of hardness are also in use in various countries.

$$= \frac{(mass\ of\ hardness\ producing\ substance)\ (chemical\ equivalent\ of\ CaCO_3)}{chemical\ equivalent\ of\ hardness\ producing\ substance}$$

$$= \frac{(mass\ of\ hardness\ producing\ substance)\ (50)}{chemical\ equivalent\ of\ hardness\ producing\ substance}$$

$$1\ ppm = \frac{1\ part\ of\ hardness}{10^6\ parts\ of\ water}$$

Relation between various units of hardness

$$1 \text{ ppm} = 1 \text{ mg/L} = 0.1° \text{ Fr} = 0.07° \text{ Cl}$$
$$1 \text{ mg/L} = 1 \text{ ppm} = 0.1° \text{ Fr} = 0.07° \text{ Cl}$$
$$1° \text{ Cl} = 1.43° \text{ Fr} = 14.3 \text{ ppm} = 14.3 \text{ mg/L}$$
$$1° \text{ Fr} = 10 \text{ ppm} = 10 \text{ mg/L} = 0.7° \text{ Cl}$$

I.5 Estimation of Hardness of Water by EDTA Method

I.5.1 Principle

Total hardness is caused by the presence of bicarbonates, chlorides and sulphates of calcium and magnesium ions. The total hardness of water is estimated by volumetric titration of water sample against ethylenediamine tetraacetic acid (EDTA) using Eriochrome Black-T (EBT) indicator. The amount of EDTA consumed by the hard water reveals the amount of calcium/magnesium ions present.

Initially EBT forms a weak EBT-Ca^{2+}/Mg^{2+}, which is a wine red coloured complex (equation (a)). On addition of EDTA solution, Ca^{2+}/Mg^{2+} ions preferably form a stable EDTA Ca^{2+}/Mg^{2+} complex with EDTA, leaving the free EBT indicator in solution which is steel blue in colour (equations (b)-(d)) in the presence of ammonia buffer (pH 10).

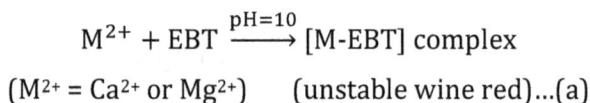

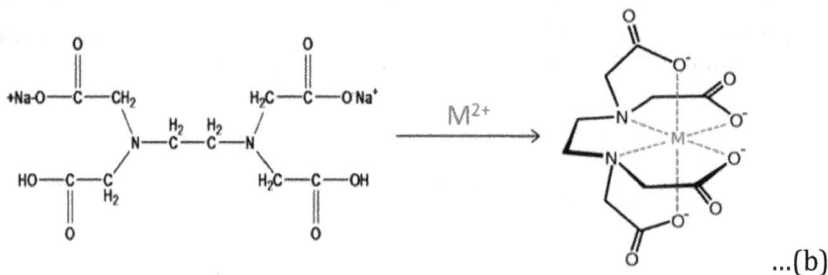

$$M^{2+} + \text{EBT} \xrightarrow{\text{pH=10}} \text{[M-EBT] complex}$$

$$(M^{2+} = Ca^{2+} \text{ or } Mg^{2+}) \qquad \text{(unstable wine red)...(a)}$$

...(b)

$$M + \text{EDTA} \rightarrow \text{[M - EDTA]...(c)}$$

[M-EBT] complex + EDTA → [M-EDTA] complex + EBT

(wine red)   (stable complex)   (blue)...(d)

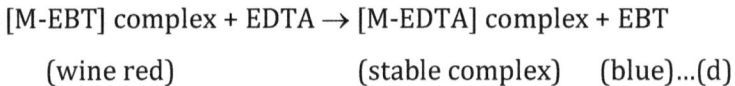

In this titration, Eriochrome Black–T [EBT] is used as indicator, which also forms a less stable complex.

Calculation of total hardness
Volume of EDTA solution consumed   = ... mL
Volume of hard water taken   = ... mL

$$\text{Total hardness} = \frac{\text{Normality of EDTA x volume of EDTA}}{\text{volume of hard water taken}} = ...\text{ppm}$$

Calculation of permanent hardness
Volume of EDTA solution consumed   = ... mL
Volume of boiled water taken   = ... mL

$$\text{Permanent Hardness} = \frac{\text{Normality of EDTA x volume of EDTA}}{\text{volume of boiled water taken}} = ...\text{ ppm}$$

Calculation of temporary hardness
Temporary hardness of the given sample of water = Total hardness - Permanent hardness = ... ppm

---

Numerical Problems based on Hardness of Water

1. Calculate the temporary and permanent hardness of water sample containing $Mg(HCO_3)_2$ = 8.3 mg/L, $Ca(HCO_3)_2$ = 17.2 mg/L, $MgCl_2$ = 8.5 mg/L, $CaSO_4$ = 12.6 mg/L.

Solution: conversion into $CaCO_3$ equivalents:

Constituent	Multiplication factor	CaCO3 equivalent
$Mg(HCO_3)_2$ = 8.3 mg/L	100/146	8.3x100/146 = 5.6849 mg/L
$Ca(HCO_3)_2$ = 17.2 mg/L	100/162	17.2x100/162 = 10.6172 mg/L
$MgCl_2$ = 8.5 mg/L	100/95	8.5x100/95 = 8.9473 mg/L
$CaSO_4$ = 12.6 mg/L	100/136	12.6x100/136 = 9.2647 mg/L

Therefore, the temporary hardness of water due to $Mg(HCO_3)_2$ and $Ca(HCO_3)_2$ = 5.6849 + 10.6172
= 16.3021 mg/L or 16.30 ppm.

Permanent hardness due to $MgCl_2$ and $CaSO_4$ = 8.9473 + 9.2647 = 18.2120 mg/L or 18.21 ppm.

2. 50 ml of a sample water consumed 14 mL of 0.01 M EDTA before boiling and 5 mL of the same EDTA after boiling. Calculate the degree of hardness, permanent hardness and temporary hardness.

Solution: 50 mL of water sample = 14 mL of 0.01 M EDTA
= (14 x 100)/50 mL of 0.01 M EDTA = 280 mL of 0.01 M EDTA
= 2 x 280 mL of 0.01 N EDTA (Molarity of EDTA = 2 x Normality of EDTA)
= 560 mL or 0.5 L of 0.01 eq. of $CaCO_3$
= 0.56 x 0.01 x 50 g $CaCO_3$ eq.
Hence, total hardness = 0.28 g or 280 mg of $CaCO_3$ eq.
= 280 mg/L or ppm.
Now, 50 mL of boiled water = 5 mL of 0.01 M EDTA
Therefore, 1000 mL of boiled water = (5 x 1000)/50 mL of 0.01 M EDTA
= 100 mL of 0.01 M EDTA
= 200 mL or 0.2 L of 0.01 N EDTA
= 0.2 x 0.01 x 50 g of $CaCO_3$ eq.
= 0.1 g or 100 mg of $CaCO_3$ eq.
Hence, permanent hardness = 100 mg/L or ppm.
Temporary hardness = 280 - 100 = 180 ppm.

3. 0.5 g of $CaCO_3$ was dissolved in HCl and the solution made up to 500 mL with distilled water. 50 mL of the solution required 48 mL of EDTA solution for titration. 50 mL of hard water sample required 16 mL of EDTA and after boiling and filtering required 10 mL of EDTA solution. Calculate the hardness.

Solution: 500 mL of SHW = 1 mg $CaCO_3$ eq.
Therefore, 1 mL SHW = 1 mg $CaCO_3$ eq.
Now, 48 mL of EDTA solution = 50/48 mg $CaCO_3$ eq.
Therefore, 1 mL of EDTA solution = 50/48 mg $CaCO_3$ eq.
50 mL hard water = 16 mL EDTA = 16 x 50/48 mg of $CaCO_3$ eq.
Therefore, 1000 mL of hard water = (16.667 x 1000)/50 = 333.332 mg $CaCO_3$ eq.
Hence, total hardness = 333.332 mg/L or 333.332 ppm.

I.6 Boiler Feed Water

A boiler is a closed vessel used to produce steam by evaporating water under pressure for industries and power houses.

I.6.1 Scale and Sludge Formation in Boilers

In boilers, since water evaporates continuously, the concentration of the dissolved salts increases. As the dissolved salts concentration reaches the saturation point, the salt precipitates out of water and settles on the inner walls of the boiler.

If the precipitation takes place in the form of loose and slimy precipitate, it is called ***sludge***.
If the precipitated salt forms a hard adhering crust or coating on the inner walls of the boiler, it is called ***scale***.

Sludge/scale is a poor conductor of heat which tends to waste a portion of heat supplied to generate steam, thus, resulting in the wastage of fuel. Sludge or scale formation seriously affects the working of the boiler as it settles in pipe connections, plug openings, guage-glass connections, thereby, causing choking of the pipes, lowering of boiler safety, decrease in efficiency and danger of explosion.
A boiler feed water should correspond to the following composition:
1. Its hardness should be below 0.2 ppm.
2. Its caustic alkalinity (due to OH-) should be in between 0.15 ppm to 0.45 ppm.
3. Its soda alkalinity (due to Na_2CO_3) should be in between 0.45 ppm to 1 ppm.

I.6.2 Treatment of Boiler Feed Water

Internal treatment involves addition of chemicals to the boiler water to either (i) precipitate the scale forming impurities in the form of sludge, which can be easily removed, or (ii) convert the impurities to soluble compounds, so that scale formation can be avoided. Important internal treatments involve:

(i) Phosphate conditioning: Different sodium phosphates like NaH_2PO_4, Na_2HPO_4 and Na_3PO_4 are added to high pressure boilers to react with the hardness forming impurities to form soft sludge of calcium and magnesium phosphates, which can be removed by blow down operation.

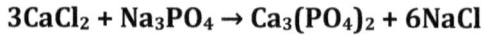

$$3CaCl_2 + Na_3PO_4 \rightarrow Ca_3(PO_4)_2 + 6NaCl$$

(ii) Colloidal conditioning: Organic substances like kerosene, tannin and agar-agar are added to form gels and loose non-sticky deposits with scale-forming precipitates, which can be easily removed by blow-down operations in low pressure boilers.

(iii) Sodium aluminate conditioning: Sodium aluminate is hydrolysed yielding NaOH and gelatinous $Al(OH)_3$. The formed NaOH reacts with magnesium salts to precipitate as $Mg(OH)_2$. The $Mg(OH)_2$ and $Al(OH)_3$ are flocculants and entrap the colloidal as well as finely divided impurities like silica in the boiler water and the loose precipitate is finally removed by blow down operation.

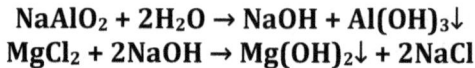

$$NaAlO_2 + 2H_2O \rightarrow NaOH + Al(OH)_3\downarrow$$
$$MgCl_2 + 2NaOH \rightarrow Mg(OH)_2\downarrow + 2NaCl$$

(iv) Calgon conditioning: Calgon, (sodium hexameta phosphate), when added to boiler water, reacts with scale forming $CaSO_4$ and forms soluble complex compound.

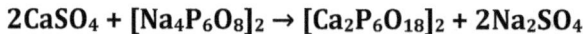

$$2CaSO_4 + [Na_4P_6O_8]_2 \rightarrow [Ca_2P_6O_{18}]_2 + 2Na_2SO_4$$

(v) Carbonate conditioning: Sodium carbonate is added to the water of low pressure boiler, whereby the scale forming $CaSO_4$ gets converted to loose sludge of $CaCO_3$, which can be easily removed by blow-down operation.

$$CaSO_4 + Na_2CO_3 \rightarrow CaCO_3 + Na_2SO_4$$

(vi) Electrical conditioning: Rotating mercury bulbs on heating by the boiling water emit electrical discharges that prevent scale formation by the particles.

External treatment includes ion exchange process, zeolite process, desalination of brackish water and reverse osmosis, among other methods.

(i) Ion exchange (IE) process: In this method, the hardness causing cations such as Ca^{2+}/Mg^{2+} are removed by exchanging with sodium ions which contributes to a small extent (non-objectionable) to the hardness of water. The process in which an ion is exchanged by another ion is called ion exchange. Both the contaminant and the exchanged substance must be dissolved and have the same type (+,-) of electrical charge (Figure I.1).

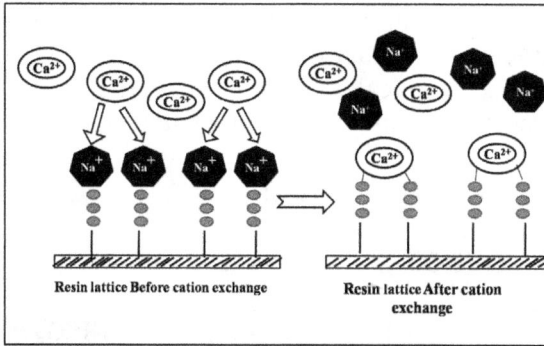

Figure I.1 Ion exchange process.

In this process involving exchange of cations, positively charged ions are exchanged with positively charged ions available on the resin surface, usually sodium. In the anion exchange process, negatively charged ions are exchanged with negatively charged ions on the resin surface, usually chloride. Various contaminants, including nitrate, fluoride, sulfate and arsenic, can be removed by anion exchange.

The ability of ion-exchange resin materials (either organic polymers or inorganic zeolites) to undergo exchange of ions previously attached and loosely incorporated into its structure by oppositely charged ions present in the surrounding solution, is called ion-exchange capacity. The clay minerals and synthetic substances (polymer resins) have high cation-exchange capacities.

The efficacy of ion exchange for water treatment can be limited by mineral scaling, surface clogging and other issues that contribute to resin fouling. Pretreatment processes such as filtration or addition of chemicals can help reduce or prevent these issues.

In ion exchange softening process, also known as the zeolite softening process, water is passed through a filter containing resin granules. In the filter, known as a softener, calcium and magnesium ions in water are exchanged with sodium from the zeolite bed. The resulting water has a hardness of 0 mg/L and must be mixed with hard water to prevent softness problems in the distributed water.

Zeolite minerals used in water softening, for example, have a large capacity to exchange sodium ions (Na^+) with calcium ions (Ca^{2+}) of hard water. Each of the individual exchange sites become saturated with prolonged use. Thus, the resin must be recharged or regenerated to restore it to its initial condition. The substances used for this process can include sodium chloride, as well as hydrochloric acid, sulfuric acid or sodium hydroxide.

The following chemical reactions show the exchange process, where X represents zeolite, the exchange material:

Removal of carbonate hardness:

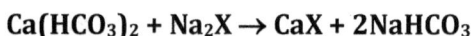

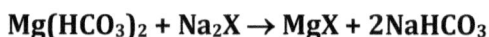

$$Ca(HCO_3)_2 + Na_2X \rightarrow CaX + 2NaHCO_3$$

$$Mg(HCO_3)_2 + Na_2X \rightarrow MgX + 2NaHCO_3$$

Removal of non-carbonate hardness:

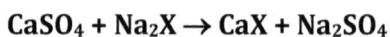

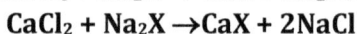

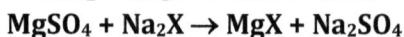

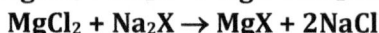

$$CaSO_4 + Na_2X \rightarrow CaX + Na_2SO_4$$
$$CaCl_2 + Na_2X \rightarrow CaX + 2NaCl$$
$$MgSO_4 + Na_2X \rightarrow MgX + Na_2SO_4$$
$$MgCl_2 + Na_2X \rightarrow MgX + 2NaCl$$

Figure I.2 and Table I.1 also show the ion exchange column and various types of ion exchange columns.

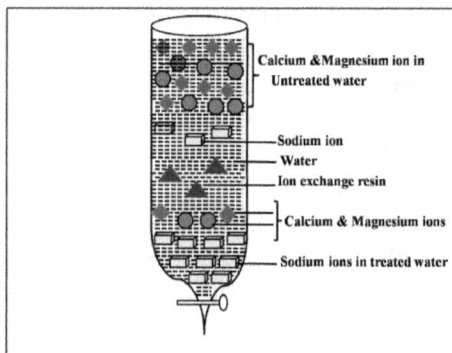

Figure I.2 Ion exchange column demonstrating replacement of Na^+ ions.

Table I.1 Types of ion exchange resins

$$NaSO_3 + NaHCO_3 \xrightarrow[\text{Exchanger}]{\text{Anion Cl}^-} NaCl + NaCl...(i)$$

$$H_2CO_3 + H_2SO_4 \xrightarrow[\text{Exchanger}]{\text{Anion OH}^-} H_2O + H_2O...(ii)$$

$$Ca(HCO_3)_2 + CaSO_4 \xrightarrow[\text{Exchanger}]{\text{Cation Na}^+} NaHCO_3 + Na_2SO_4...(iii)$$

$$Ca(HCO_3)_2 + CaSO_4 \xrightarrow[\text{Exchanger}]{\text{Cation H}^+} H_2CO_3 + H_2SO_4...(iv)$$

$$Ca(HCO_3)_2 \xrightarrow[\text{Exchanger}]{\text{Cation H}^+} H_2CO_3...(v)$$

- **Conventional softening – Process (iii)**
- **Dealkalization by split stream softening – blending effluents from (iii) and (iv)**
- **Dealkalization by anion exchanger – Process (i) preceded by (iii)**
- **Dealkalization by weak acid cation exchanger followed by conventional softening process (v), followed by (iii)**
- **Demineralizing – combination of (iv) and (ii)**

Polymer based cation-exchange resins have exchangeable sodium ions or protons, as given in Table 1.1. The resin with highly reactive $-SO_3H-$ is considered to be strong cation exchanger and one with $-COO-$ functional groups is the weak cation exchanger. Some of the commercially available exchange resins are sulfonated coal, polystyrene-beads, phenolic resins and acrylic resins (as shown in Figure I.3).

Anion exchange materials (Figure I.4) are also classified as weak base or strong base exchangers depending on the type of functional groups. Weak base resins act as acid adsorbers, efficiently removing strong acids such as sulfuric and hydrochloric. However, these will not remove carbon dioxide or silica.

(ii) Desalination of brackish water: water having higher saline content (500-15,000 ppm of NaCl) than fresh water and less content than sea water, is known as brackish water. It results from mixing of the seawater with fresh water in estuaries or civil engineering projects such as dikes and from flooding of coastal marshland to produce brackish water pools for the purpose of freshwater prawn farming.

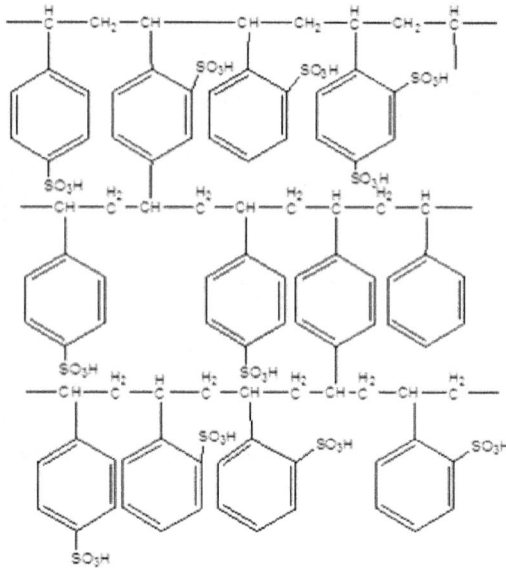

Figure I.3 Sulphonated polystyrene based cation exchange resin.

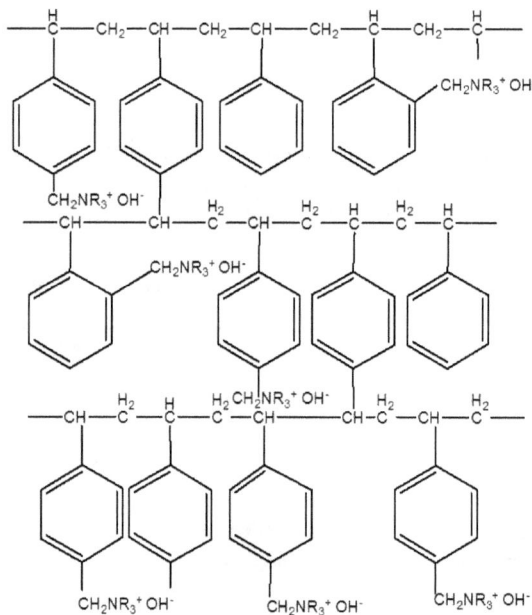

Figure I.4 Quaternary ammonium salt of polystyrene based anion exchange resin.

The desalination process separates water from its dissolved salts either by distillation or by membrane filtration. In this process, water is separated into two parts, as shown in Figure I.5, one that has a low concentration of salt (treated water or product water) and the other one with much higher concentration of salt than the feed water, referred to as brine water or 'concentrate' or reject water.

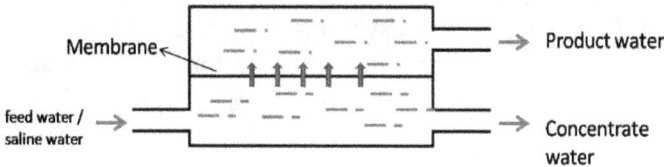

Figure I.5 The desalination process.

The two major techniques employed in desalination are thermal and membrane technologies. The major desalination processes based on these are:

Thermal technology	Multi-stage flash distillation Multi effect distillation Vapour compression distillation
Membrane technology	Electrodialysis Reverse electrodialysis Reverse osmosis

The process of multi stage flash distillation is depicted in Figure I.6. Sea water is passed through tube 'd' which acts as a condenser.

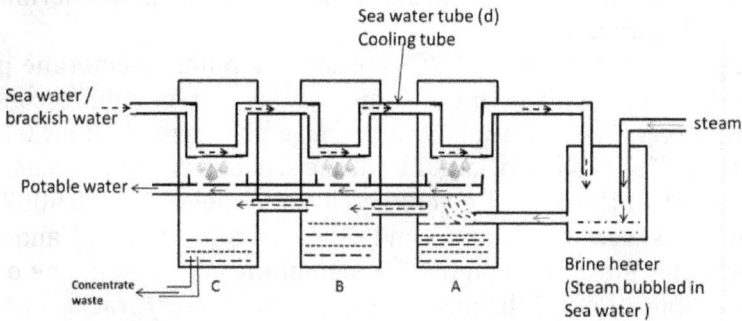

Figure I.6 Multi stage flash distillation process.

The sea water or brackish water reaches brine heater, where steam is bubbled into it. The sea water evaporates with steam and condenses at tower A. The uncondensed sea water and steam move to towers B and C.

(iii) Reverse osmosis (RO): One of the first methods developed for desalination of brackish/saline water is membrane separation involving reverse osmosis. RO is a process using a mechanism different from distillation or ion exchange technique. When the pressurized feed water flows across a membrane surface, portion of water permeates through the membrane leaving salts rich water (as shown in Figure I.7). This type of system is called cross-flow filtration.

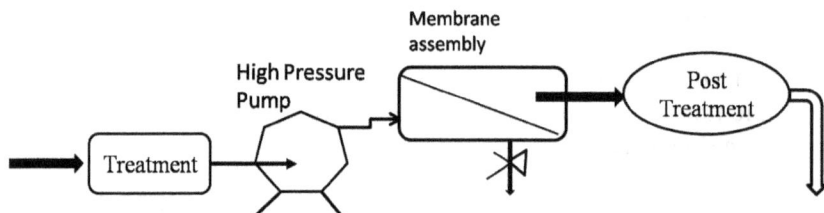

Figure I.7 Cross-flow filtration.

Osmosis is a phenomenon where pure water flows from a dilute solution through a semi-permeable membrane to a higher concentrated solution. As a pressure is greater than the osmotic pressure at the high concentration is applied, the direction of water flow through the membrane can be reversed. This is called reverse osmosis (abbreviated RO). Note that this reversed flow produces pure water from the salt solution, since the membrane is not permeable to salt (Figure I.8).

The basic equipment of RO consists of a pump, membrane pressure housing, semi-permeable membrane and plumbing connections. Reverse osmosis membranes have pore size of about 0.0005 microns. This is slightly larger than the size of a water molecule, but smaller than the size of a sodium chloride molecule at 0.0007 micron. Thus, sodium chloride and other salts are blocked and only pure water flows through the RO membrane. As a result, one of the main applications of RO membranes is in seawater/brackish desalination. Seawater RO has a conversion rate of 35% to 40% and 90% for brackish water.

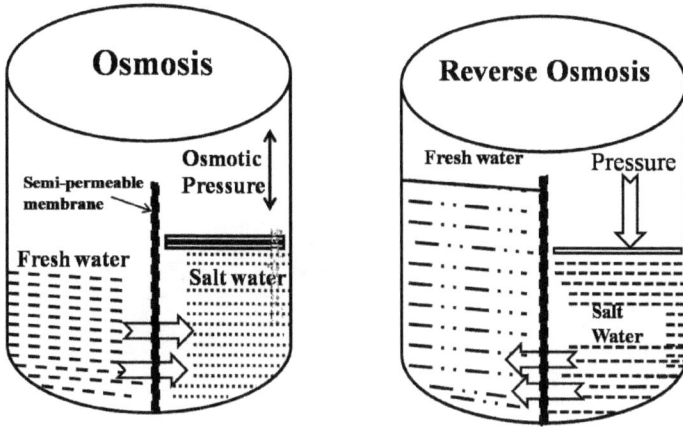

Figure I.8 The RO process.

SURFACE CHEMISTRY AND CATALYSIS

II.1 Adsorption

Adsorption is defined as the deposition of atoms, molecules or ions onto the surface of solid or liquid. The atoms, ions and molecules getting adsorbed on the surface are known as **adsorbate** and the surface on which adsorption occurs is known as **adsorbent**. Common examples of adsorbents are clay, silica gel, colloids, metals, etc. The adsorption is different from absorption which refers to the molecules entering into the bulk of the solid material.

Adsorption	Absorption
The higher concentration of any molecular species on a surface of solid (or liquid) than in the bulk of same is known as adsorption	Penetration of molecules throughout the solid or liquid, in uniform concentration
Surface phenomenon	Bulk phenomenon
Influenced by temperature and pressure	It is not influenced by temperature or pressure
Examples: (i) Water vapour is absorbed by anhydrous calcium chloride and adsorbed on to charcoal. (ii) Ammonia is absorbed by water and adsorbed on to charcoal. (iii) bloating paper absorbs water.	

The driving force for adsorption to take place is surface energy (forces). The surface energy may be defined as the excess energy acting at the surface of a material compared to bulk. As we know, the residual forces acting along the surface of a liquid give rise to surface tension. Similar forces also exist in solids. These forces have a tendency of attracting molecules of other species with which they are brought into contact.

Exothermic nature of adsorption: Adsorption decreases residual forces, thereby, resulting in a decrease of surface energy, which appears in the form of heat. The amount of heat evolved when one mole of any gas or vapour is adsorbed on a solid surface is called enthalpy of adsorption or ***molar heat of adsorption.***

II.2 Types of Adsorption

Adsorption is classified into two types based on the nature of forces which hold adsorbate and adsorbent as
- Physical adsorption (physisorption)
- Chemical adsorption (chemisorption)

Physisorption	Chemisorption
Weak van der Waals forces of attraction	Adsorbent-adsorbate chemical bond is formed
Physical adsorption	Chemical adsorption
Reversible	Irreversible
Multilayer	Monolayer
20-40 kcal of heat of adsorption	40-400 kcal of heat of adsorption
Adsorption decreases with rise in temperature above boiling point	Adsorption can occur at higher temperature
Rate of adsorption increases with pressure near saturation, multilayer is formed	Decreases with rise in pressure or concentration, rate decreases near saturation
Amount adsorbed on the surface	Amount adsorbed is characteris-

is more a function of adsorbate than adsorbent	tic of both adsorbent and adsorbate
Little activation energy change	Involves appreciable activation energy
Equilibrium attained rapidly	Equilibrium requires time
Not specific in nature	Highly specific in nature

II.3 Adsorption of Gases on Solids

As a solid surface is exposed to a gas at pressure p, gas molecules adsorb on the solid surface, weight of the solid increases and the pressure of the gas decreases with time. After a time t, the pressure p and weight become constant because a dynamic equilibrium is reached. The amount of gas adsorbed is experimentally determined by gravimetry (weight gained by solid), volumetry (decrease in gas pressure) and the heat evolved.

The important factor pertaining to adsorption is surface coverage and rate of adsorption. The fractional coverage θ is defined as

$$\theta = \frac{\text{number of adsorption sites occupied}}{\text{number of adsorption sites available}}$$

II.3.1 Factors Influencing Adsorption

(i)   Surface area: directly proportional to adsorbed quantity; higher the surface area higher the adsorbed gas molecules.

(ii)   Finely divided metals and porous substances with large surface area have higher adsorption capacity.

(iii) Activated surface: Removal of impurities and cleaning the surface of adsorbent is called activation. Activation is achieved by heating in vacuum or in the presence of an inert gas. An activated adsorbent has higher adsorption capacity.

(iv) Nature of the gas: easily liquifiable gases such as HCl, NH_3, Cl_2 are adsorbed more easily.

(v)   Nature of adsorbent

(vi)   Pressure: With increase in pressure, physical adsorption increases; effect of pressure on chemical adsorption is negligible.

(vii) Temperature: Low temperature favours physical adsorption and high temperature favours chemical adsorption.

II.4 Adsorption Isotherms

Adsorption process is better understood by plotting amount of adsorbate deposited on the adsorbent surface as a function its pressure or concentration at constant temperature, known as adsorption isotherm.

These adsorption isotherms explain the variation of magnitude of adsorption with pressure at given temperature. The equilibrium relationship between amount of adsorbate adsorbed and the amount of unadsorbed adsorbate in solution or gas is described by adsorption isotherms.

Considering x as mass of adsorbate and M as mass of adsorbent, the variation of adsorption with pressure at a given constant temperature can be expressed as shown in Figure II.1.

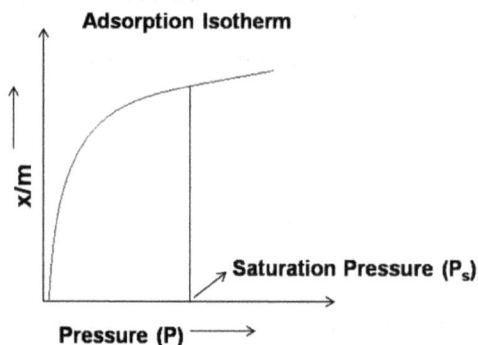

Figure II.1 Variation of adsorption with pressure.

As shown in Figure II.1, the rate of adsorption increases with pressure. After reaching a limiting value, the rate of adsorption remains constant and the corresponding pressure is called equilibrium pressure. At this stage, the rate of adsorption and rate of desorption are same.

II.4.1 Freundlich Adsorption Isotherm

In 1909, Freundlich developed an equation to explain the variation of adsorption of the amount of gas (adsorbate) adsorbed by unit mass of solid adsorbent with pressure at constant temperature. This equation is known as Freundlich adsorption isotherm.

$$\frac{x}{M} = kP^{\frac{1}{n}}$$

where x is mass of adsorbate, M is mass of adsorbent, P is the pressure, k and n are constants whose values depend upon adsorbent and adsorbate at a particular temperature.

Figure II.2 depicts the typical Freundlich Isotherm plotted x/m against pressure 'p'. The value of x/m increases with increase in 'p', thus, forming a straight line. After certain point, x/m increases slowly, but not proportionate to 'p', thus, forming a curve. This curve is called as Freundlich isotherm curve. The Freundlich Isotherm explains relationship of adsorption at low pressure appropriately, but has limitations in accurately predicting value of adsorption at higher pressure.

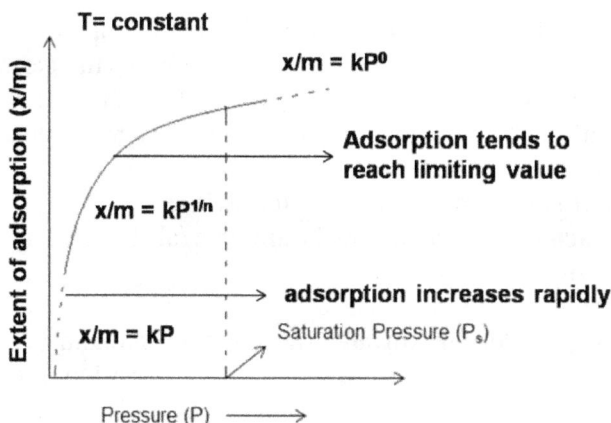

Figure II.2 Freundlich isotherm curve.

Taking the logarithms of first equation,

$$\frac{\log (x)}{M} = \log k + \frac{1}{n}\log P$$

Hence, if a graph of log(x)/m is plotted against log(P), it will result in a straight line, as shown in Figure II.3.

From this graph, the value of slope is equal to 1/n and the value of intercept is equal to log k. Thus, if the graph of log(x)/m against log(P) is a straight line, it can be assured that the Freundlich adsorption isotherm is satisfied for the system.

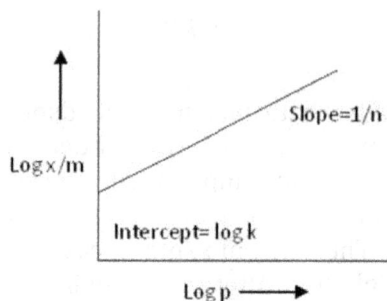

Figure II.3 Plot of log(x)/m against log(P).

II.4.2 Langmuir Adsorption Isotherm

Irving Langmuir derived an isotherm for the adsorption of gas on-solids, known as Langmuir adsorption isotherm. This isotherm was derived based on some assumptions, one of which is that dynamic equilibrium exists between adsorbed gaseous molecules and the free gaseous molecules.

Langmuir isotherm assumes the following:
i) The surface of the adsorbent is uniform/all the adsorption sites are equivalent.
ii) Adsorbed molecules do not interact.
iii) All adsorption occurs through the same mechanism.
iv) Monolayer is formed at the maximum adsorption, molecules of adsorbate do not deposit on molecules already adsorbed, but adsorb only on the free surface of the adsorbent.

Langmuir suggested that adsorption takes place through this mechanism:

$$A(g) + B(s) \underset{\text{Desorption}}{\overset{\text{Adsorption}}{\rightleftharpoons}} AB$$

where, A(g) is unadsorbed gaseous molecule, B(s) is unoccupied metal surface, AB is adsorbed gaseous molecule, k and k^{-1} are rate constants of adsorption and desorption reaction.

Based on the above equation, Langmuir derived the relationship between the number of active sites of the surface undergoing adsorption with respect to gas (adsorbate) pressure. This equation is called Langmuir equation.

$$\theta = \frac{KP}{1 + KP}$$

where, θ is the number of surface sites covered with gaseous molecules, P is the pressure and K is the equilibrium constant obtained from the rate constants for adsorption and desorption.

$$K = \frac{k}{k^{-1}}$$

The limitation of Langmuir adsorption isotherm is that it is valid at low pressure only. At lower pressure, KP is negligible (very small), hence 1+KP = 1. Thus, the Langmuir equation becomes

$$\theta = KP$$

II.5 Contact Theory

II.5.1 Langmuir-Hinshelwood-Hougen-Watson Mechanism

This mechanism was first suggested by Irving Langmuir in 1921, and further developed by Cyril Hinshelwood (1926), thus, sometimes termed as Langmuir-Hinshelwood kinetics. Hougen and Watson (1943) developed a similar approach and popularized the Langmuir-Hinshelwood kinetics.

The LHHW approach assumes that all active sites are energetically uniform and, upon adsorption, adsorbed species do not interact with each other. The species adsorption restricts itself to only monolayer and the rate of adsorption is proportional to the concentration of the unoccupied (empty) active sites and the partial pressure of the component in the gas phase (Figure II.4).

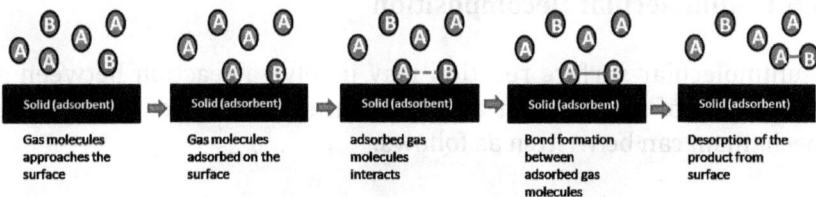

Figure II.4 Langmuir Hinselwood mechanism.

The steps involved in the process are as follows:

Adsorption: Here the reactant molecules (A, B) are adsorbed on the solid adsorbent by strong chemical bond or weak van der Waals bond.

Activated complex formation: The adjacent adsorbate molecules form weak bond (A-B) and the activated complex is formed.

Decomposition: The bond between A and B gets strengthened while the bond between A-B and adsorbent gets weakened. This is called decomposition.

Desorption: From the decomposed activated complex, the final product is released. This is known as desorption.

II.5.2 Ridal-Eley Mechanism

In this process, any one of the reactant molecules is adsorbed on the solid adsorbent by strong chemical bond or weak van der Waals bond. Adsorbed molecule interacts with an unadsorbed molecule. The bond between A and B gets strengthened, while the bond between A and adsorbent weakens. From the decomposed activated complex, the final product is released (Figure II.5).

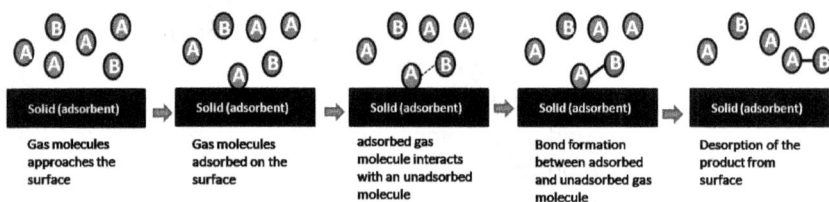

Figure II.5 Ridel-Eley mechanism.

II.6 Kinetics of Surface Reactions

II.6.1 Unimolecular Decomposition

A unimolecular surface reaction may involve a reaction between a molecule A of the reactant and a vacant site S on the surface. The mechanism can be written as follows:

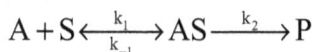

$$A + S \underset{k_{-1}}{\overset{k_1}{\longleftrightarrow}} AS \overset{k_2}{\longrightarrow} P$$

where,

A = reactant

S = vacant site on the surface

P = product

AS = reactant-substrate complex

k_1 = rate constant for the forward reaction

k_{-1} = rate constant for the breakdown of the AS to substrate

k_2 = rate constant for the formation of the products

If r is proportional to the fraction Θ of the surface covered,

$$r = k_2\Theta ...(1)$$

where, r is the rate of the reaction.

Applying steady-state approximation for the concentration of AS:

Rate of formation of [AS] = Rate of decomposition of [AS]

$$k_1[A][S] = k_2[AS] + k_{-1}[AS]$$

$$r = \frac{d[AS]}{dt} = k_1[A][S] - k_2[AS] - k_{-1}[AS] = 0 ...(2)$$

The active sites on the surface (C_s) the concentration [S] of the vacant sites on the surface is equal to $(1 - \Theta)$. Thus,

$$[S] = C_s(1 - \Theta) ...(3)$$

From (2),

$$[AS] = \frac{k_1[A][S]}{k_2+k_{-1}} ...(4)$$

Also the concentration of AS on the surface, viz. [AS] is given by

$$[AS] = C_s\Theta ...(5)$$

From eqs. (3), (4) and (5), we get

$$\Theta = \frac{k_1[A]}{k_1[A] + k_{-1}+k_2} ...(6)$$

Substituting the value of Θ in eq. 1, we get

$$r = \frac{k_1 k_2 [A]}{k_1 [A] + k_{-1} + k_2} \quad ...(7)$$

(7) may be rewritten in the form

$$\frac{1}{r} = \frac{1}{k_2} + \frac{k_{-1} + k_2}{k_1 k_2 [A]} \quad ...(8)$$

According to eq. 8, a plot of $1/r$ versus $1/[A]$ would result in a straight line with intercept equal to $1/k_2$ and slop equal to $(k_{-1} + k_2)/k_1 k_2$. For gaseous adsorbates, the concentration is conveniently expressed in partial pressure. Thus, eqs. 6 and 7 can be written as

$$r = \frac{k_1 k_2 P_A}{k_1 P_A + k_{-1} + k_2} \text{ (or) } \frac{1}{r} = \frac{1}{k_2} + \frac{k_{-1} + k_2}{k_1 k_2 P_A}$$

This equation is Langmuir–Hinshelwood mechanism.

Case I
If the rare of decomposition is very large in comparison with the rate of adsorption and desorption, $k_2 >>> (k_1[A] + k_{-1})$. So eq. 7 reduces to

$r = k_1[A]$, indicating first order reaction with respect to A.

Case II
If the rare of decomposition is very small compared with the rate of adsorption and desorption, $(k_1[A] + k_{-1}) >>> k_2$. So eq. 7 reduces to

$r = \frac{k_1 k_2 [A]}{k_1 [A] + k_{-1}}$, indicating zero order reaction (Figure II.6).

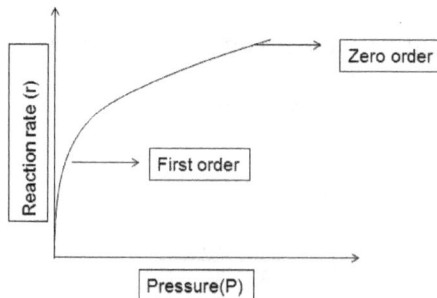

Figure II.6 Representation of the zero order and first order reactions.

II.7 Applications of Adsorption for Pollution Abatement

Generally, adsorbents are widely used in pollution control for the adsorption of pollutants such as dyes, pesticides, heavy metals, along with solvent recovery and air purification, etc. Depending on the adsorbent type applied, organic substances as well as inorganic ions can be removed from the water (Table II.1).

Table II.1 Adsorbents and their applications

Adsorbent	Objective	Application
Activated carbon	Removal of dissolved organic matter, organic micronutrients.	Drinking water, ground water and aquarium water treatment
Aluminium oxide	Removal of arsenic	Drinking water treatment and urban waste water
Iron oxide	Removal of phosphate	treatment
Polymeric adsorbents	Removal of specific chemicals, heavy metals	Industrial waste water treatment

In industrial wastewater treatment, adsorption processes is used for removal or recycling of organic substances. Activated carbon is an appropriate adsorbent for the removal of organic species from the wastewater, polymeric adsorbents are used to recycle valuable chemicals. Figures II.7 and II.8 represent two examples.

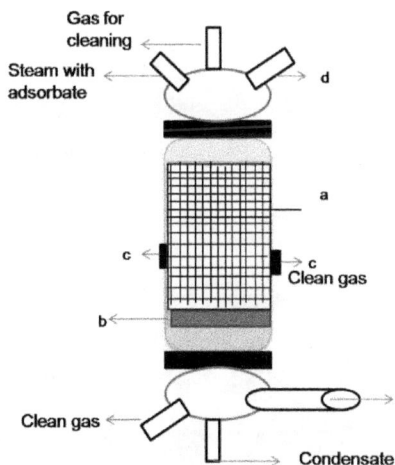

a. steel vessel
b. supporting and distributing grating
c. duct for removal of adsorbent
d. duct for loading adsorbent

Figure II.7 Fixed bed adsorber for treatment of air.

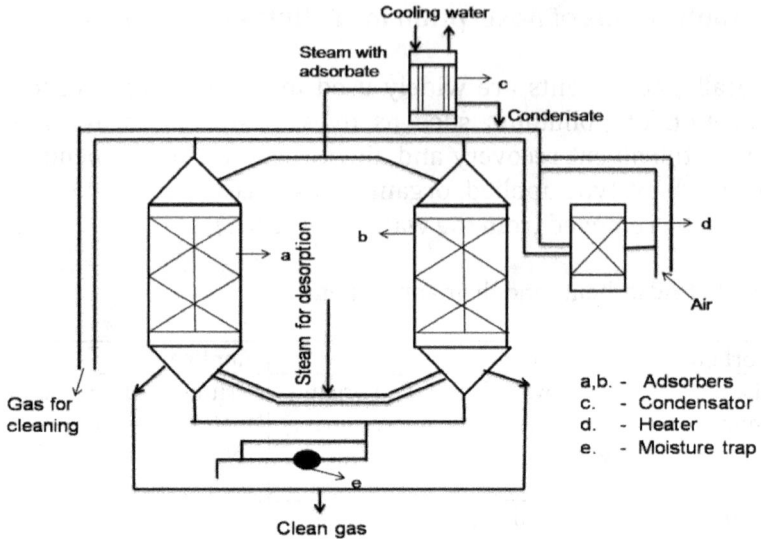

Figure II.8 Two column adsorption system for air purification.

II.8 Types of Adsorbents

Different types of conventional and non-conventional adsorbents have been used in the laboratory as well as for industrial purposes. Some of these are:

Conventional adsorbents

- activated carbon
- carbon molecular sieves (CMS)
- carbonized polymers and resins
- bone charcoal
- polymeric adsorbents
- silica gel
- activated alumina
- clay minerals
- zeolites

Non-conventional adsorbents

- adsorbent from industrial wastes

- adsorbent from coal
- adsorbent from agricultural wastes as well as agricultural byproducts
- peat
- oxides and related materials as adsorbents
- river sediments as adsorbent
- adsorbent from bio-resources
- adsorbent generated from biopolymers such as chitosan flakes
- adsorbents from various other materials

II.9 Catalysis

Catalyst is referred to substances, a small amount of which alters the rate (speed) of a chemical reaction, without itself undergoing any change in mass and composition at the end of the reaction. The process of altering the rate of a chemical reaction by the presence of catalyst is called **catalysis**. Theoretically, the catalyst lowers the activation energy E_a of a reaction, thereby, increasing the rate of the reaction, as shown in Figure II.9. Figure II.10 also demonstrates the applications of catalysis.

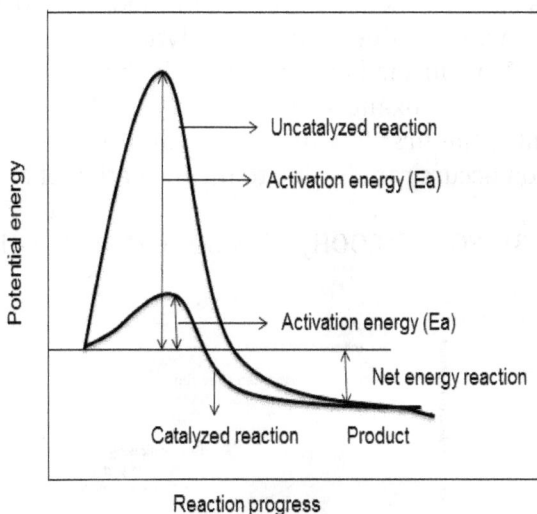

Figure II.9 Lowering of the activation energy E_a of a reaction due to the presence of a catalyst.

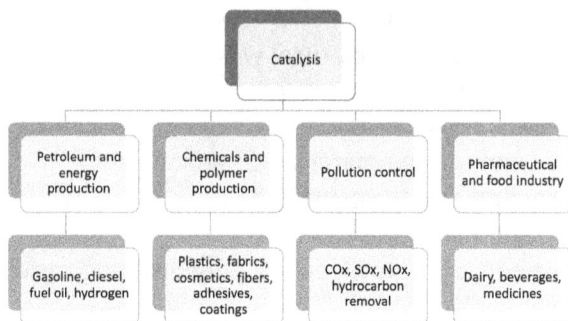

Figure II.10 Applications of catalysis.

Autocatalysis: In some reactions, one of the products formed in the course of reaction itself accelerates the reaction. Such a species is known as autocatalyst and the phenomenon is called autocatalysis. For instance, RCOOH acid formed in the below reaction accelerates the hydrolysis of ester.

$$RCOOR' + H_2O \rightarrow RCOOH + R'OH$$

In such autocatalytic processes, the initial rate of the reaction is very low. As soon as the products are formed, the rate of the reaction rapidly increases due to auto catalytic process (Figure II.11). For example, Mn^{2+} formed as product in the below reaction accelerates the oxidation of oxalic acid by $KMnO_4$. In the titration of oxalic acid by $KMnO_4$, the first few drops take appreciable time, then the decolorization occurs rapidly due to the formation of Mn^{2+} ions.

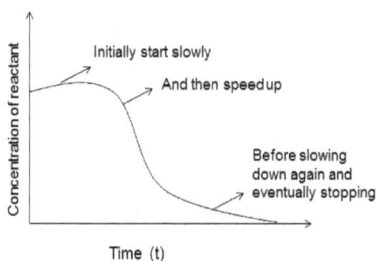

$$2KMnO_4 + 3H_2SO_4 + 2(COOH)_2 \rightarrow K_2SO_4 + 2MnSO_4 + 4H_2O + 4CO_2$$

Figure II.11 Autocatalysis process.

II.10 Characteristics of Catalysts (Criteria)

1. Catalyst remains unchanged in amount and composition at the end of the reaction, but the physical state may change. For example, coarsely grained MnO_2 becomes finely powered after the reaction.

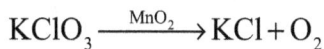

$$KClO_3 \xrightarrow{MnO_2} KCl + O_2$$

2. Catalyst cannot initiate a reaction, but can only decrease or increase the rate of the reaction.

3. Only a small amount of catalyst is sufficient to bring about appreciable change in the rate of the reaction. For example, one mole of Pt is enough to catalyze 10^8 L of H_2O_2 decomposition.

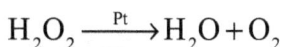

$$H_2O_2 \xrightarrow{Pt} H_2O + O_2$$

4. Catalyst does not alter the position of equilibrium in a reversible reaction.

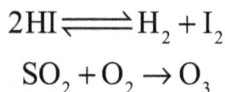

$$2HI \rightleftharpoons H_2 + I_2$$
$$SO_2 + O_2 \rightarrow O_3$$

Platinized asbestos as catalyst:

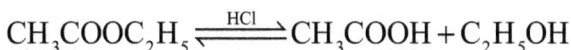

$$CH_3COOC_2H_5 \underset{}{\overset{HCl}{\rightleftharpoons}} CH_3COOH + C_2H_5OH$$

5. Catalyst is specific in its action. For example, in the blow reaction, MnO_2 can catalyze potassium chlorate but not potassium perchlorate or potassium nitrate ($KHClO_3$ or KNO_3). However, the transition metals, Fe, Co, Ni, Pt and Pd can catalyze reactions of various types.

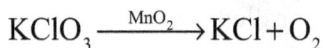

$$KClO_3 \xrightarrow{MnO_2} KCl + O_2$$

6. The catalyst does not alter the nature of the products of the reaction. For example, under suitable conditions $N_2 + H_2$ invariably forms ammonia whether catalyst used or not.

$$N_2 + H_2 \rightarrow NH_3$$

Similarly, decomposition of potassium perchlorate gives KCl and O_2 even in the uncatalyzed reaction.

$$KClO_3 \longrightarrow KCl + O_2$$

However, there are some exceptions, such as the chlorination of toluene in the presence of sunlight and $FeCl_3$:

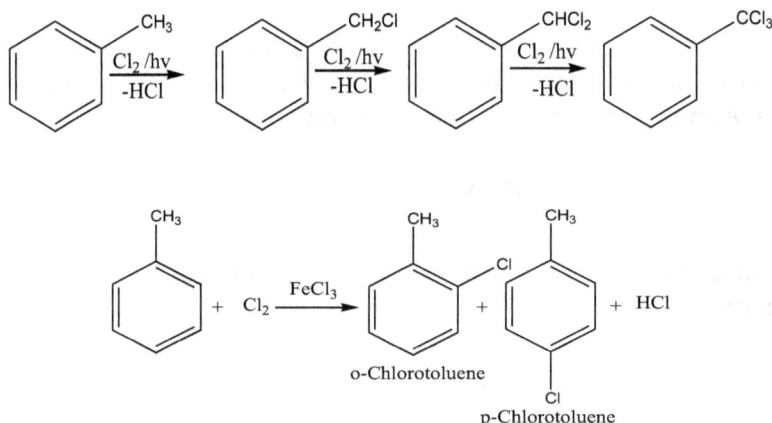

o-Chlorotoluene

p-Chlorotoluene

7. A catalyst is most active at a particular temperature called optimum temperature.
Examples: enzyme catalyst, colloidal catalyst, etc.

8. Catalyst is poisoned by certain substances. Certain impurities retard the velocity of chemical reaction by deactivating the catalyst .

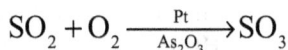

$$SO_2 + O_2 \xrightarrow[As_2O_3]{Pt} SO_3$$

9. **Catalytic promotors or activators**: The addition of small amount of foreign substances called promoters or activators, which are though not catalytically active themselves, can enhance the activity of the catalyst.

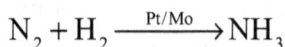

$$N_2 + H_2 \xrightarrow{Pt/Mo} NH_3$$

(i) For example, in Haber's process for the synthesis of ammonia, traces of molybdenum increases the activity of finely divided iron which acts as a catalyst.

(ii) In the manufacture of methyl alcohol from water gas, chromic oxide is used as a promoter with the catalyst zinc oxide.

Catalytic poisons: Substances which destroy the activity of the catalyst by their presence are known as **catalytic poisons.**

(i) For example, the presence of traces of arsenious oxide in the reacting gases reduces the activity of platinized asbestos which is used as catalyst in contact process for the manufacture of sulphuric acid.

(ii) The activity of iron catalyst is destroyed by the presence of hydrogen sulphide or carbon monoxide in the synthesis of ammonia by Haber's process.

(iii) The platinum catalyst used in the oxidation of hydrogen is poisoned by carbon monoxide.

10. When a catalyst accelerates the speed of the reaction, it is called a **positive catalyst.**

For example,

$$KClO_3 \xrightarrow{\ MnO_2\ } KCl + O_2$$

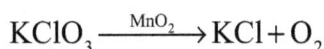

If the catalytic substance retards the chemical reaction, it is called a **negative catalyst.**

For example,

$$2CHCl_3 + O_2 \xrightarrow{\ ROH\ } 2Cl-CO-Cl + 2HCl$$

$$2H_2O_2 \xrightarrow{\ Glycerine\ } H_2O + O_2$$

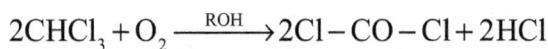

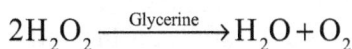

II.11 Types of Catalysis

Homogeneous catalysis: Catalyst and reacting substances are in same phase (liquid or gas). In a homogeneous catalytic reaction, the catalyst is in the same phase as the reactants.

For example, formation of sulphur trioxide from sulphur dioxide and oxygen, in the presence of NO gas as catalyst:

$$2SO_2(g) + O_2(g) \xrightarrow{\text{NO(g)}} 2SO_3(g)$$

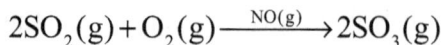

Other examples:
1. Homogeneous catalysis in gaseous phase: decomposition of acetaldehyde in the presence of iodine vapours as catalyst

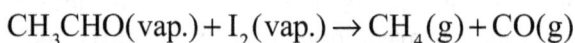

$$CH_3CHO(vap.) + I_2(vap.) \rightarrow CH_4(g) + CO(g)$$

2. Homogeneous catalysis in solution phase: hydrolysis of ester in the presence of an acid as catalyst

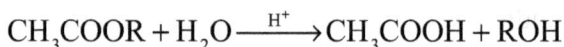

$$CH_3COOR + H_2O \xrightarrow{\text{H}^+} CH_3COOH + ROH$$

Heterogeneous catalysis: The catalyst and the reacting substances are present in different phase. Reactions of liquid or gases in the presence of solid catalysts are the typical examples. An example is the contact process for manufacturing sulphuric acid, in which the sulphur dioxide and oxygen are passed over a solid vanadium oxide catalyst producing sulphur trioxide. Several hydrocarbon transformation reactions such as cracking, reforming, dehydrogenation, isomerization also involve heterogeneous catalysis.

Heterogeneous catalyst offer several intrinsic advantages over their homogeneous counterparts, ease of product separation and catalyst reuse, along with process advantages through reactor operation in continuous flow configuration.

Examples:
1. Contact process involving synthesis of sulphur trioxide from sulphur dioxide using vanadium oxide (V_2O_5) as catalyst. Here, the reactants are in gas phase and catalyst is in solid phase (Figure II.12).

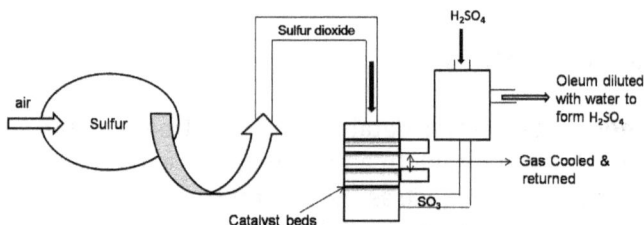

Figure II.12 Contact process.

2. Habers process: The synthesis of ammonia from hydrogen and nitrogen gases using solid catalysts based on iron promoted with K_2O, CaO, SiO_2 and Al_2O_3 (Figure II.13).

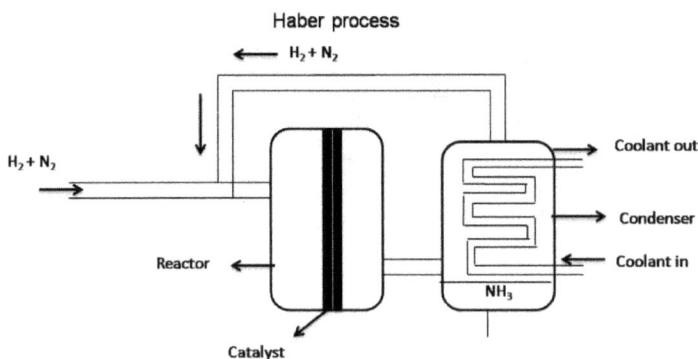

Figure II.13 Habers process

3. Ostwald process: Ammonia is oxidized by heating with oxygen in the presence of a catalyst such as platinum with 10% rhodium, to form nitric oxide. Subsequently, the formed nitrogen dioxide is absorbed in water, thus, forming forms nitric acid.

II.12 Acid Base Catalysis

Acid base catalysis is the acceleration of a chemical reaction by the addition of an acid or a base, while the acid or base itself not being consumed in the reaction. Many reactions are catalyzed by both acids and bases. The mechanism of acid and base-catalyzed reactions is explained by Brønsted-Lowry concept of acids and bases.

II.12.1 Brønsted - Lowry Acid-Base Theory

Acid: Hydrogen containing species able to donate a proton
Base: Species capable of accepting a proton

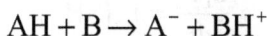

$$AH + B \rightarrow A^- + BH^+$$

In acid catalysis, a Bronsted acid donates a proton to a substrate allowing it to attain a high energy transition state, which facilitates its conversion to the product. In the case of a base catalyzed reaction, a Bronsted base extracts a proton from the substrate leading to the formation of a transition state intermediate.

Acid Catalyzed Ester Hydrolysis :

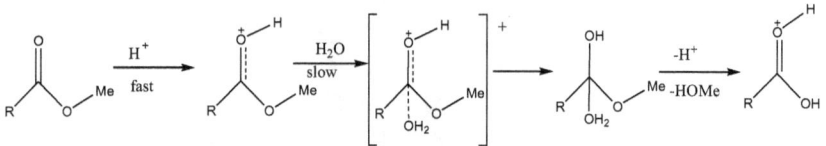

Base Catalyzed Ester Hydrolysis :

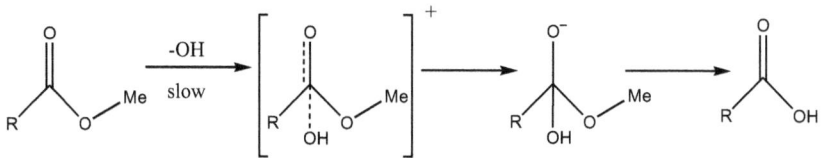

Acid catalysis involves an equilibrium reaction in which a transfer of proton takes place from an acid to a substrate 'S'. The protonated substrate then reacts to generate product and proton.

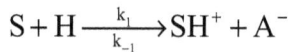

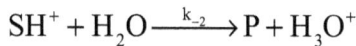

$$S + H \xrightarrow[k_{-1}]{k_1} SH^+ + A^-$$

$$SH^+ + H_2O \xrightarrow{k_2} P + H_3O^+$$

The rate of appearance of product is given by

$$\frac{d[P]}{dt} = k_2[SH^+][H_2O]$$

Since water is present in excess, the equation can be written in simple form as

$$\frac{d[P]}{dt} = k_2[SH^+]$$

The concentration of SH+ can be determined by applying steady sate approximation.

Base catalysis involves transfer of a proton from the substrate molecule to the base.

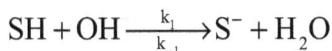

$$SH + OH \xrightarrow[k_{-1}]{k_1} S^- + H_2O$$

$$S^- \xrightarrow{k_2} P + OH^-$$

It has been observed that not only H^+ ions, but all Bronsted acids (proton donors) can be used as acid catalysts. Some common acid catalysts are, H+, undissociated acid (CH_3COOH), cations of weak bases (NH_4^+) and water (H_3O^+). Also, not only OH^- ions, but all Bronsted bases (proton acceptors) cause base catalysis.

II.12.2 Applications of Acid Base Catalysis

Acid catalysis is employed in a large number of industrial reactions, among them the conversion of petroleum hydrocarbons to gasoline and related products is very common. Such reactions include decomposition of high-molecular-weight hydrocarbons (cracking) using alumina–silica catalysts (Brønsted-Lowry acids), polymerization of unsaturated hydrocarbons using sulfuric acid or hydrogen fluoride (Brønsted-Lowry acids) and isomerization of aliphatic hydrocarbons using aluminum chloride (a Lewis acid). Among industrial applications of base-catalyzed reactions is the reaction of diisocyanates with polyfunctional alcohols in the presence of amines, used in the manufacture of polyurethane foams.

II.13 Catalytic Converter

A **catalytic converter** is a device used to reduce the toxicity of emissions from an internal combustion engine. It converts toxic gases and pollutants in the exhaust into less-toxic pollutants by catalyzing a redox reaction (an oxidation and a reduction reaction). The catalytic converter converts toxic carbon monoxide, hydrocarbons, nitric oxide (NO), nitrogen dioxide (NO_2) and nitrous oxide (N_2O) into less toxic carbon dioxide, water and nitrogen. For instance, oxidation of carbon monoxide to carbon dioxide occurs as:

$$2CO + O_2 \rightarrow 2CO_2$$

Oxidation of hydrocarbons (unburnt and partially burnt fuel) to carbon dioxide and water:

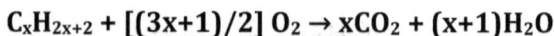

$$C_xH_{2x+2} + [(3x+1)/2]\ O_2 \rightarrow xCO_2 + (x+1)H_2O$$

Reduction of nitrogen oxides to nitrogen (N_2):

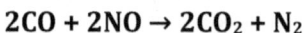

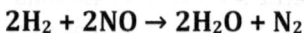

$$2CO + 2NO \rightarrow 2CO_2 + N_2$$
$$hydrocarbon + NO \rightarrow CO_2 + H_2O + N_2$$
$$2H_2 + 2NO \rightarrow 2H_2O + N_2$$

The most commonly used metallic catalysts as catalytic converters are palladium, platinum and aluminium oxide.

II.14 Solid Acid Catalyst

Solid acid catalyst (SAC) is a solid material capable of donating protons or accepting electrons, and can act as an acid to catalyze chemical reactions. SAC are alternatives to mineral acids such as HCl, HNO_3 and H_2SO_4. Solid acid catalysts are generally categorized by their Bronsted and or Lewis acidity, strength and number of sites and morphology of the support.

A wide range of liquid phase industrial reactions rely on inorganic acids. In these reactions, final isolation of the product requires aqueous quenching and neutralization steps to remove the acid, resulting in large quantities of hazardous waste, the disposal of which often outweighs the value of the product.

II.15 Enzyme Catalysis

Enzymes are biological macromolecules, present in plants, animals and microorganisms that catalyze chemical reactions occurring in living organisms. Chemically, all enzymes are protein in nature and have complex organic structure with high molecular weight. The reaction catalyzed by enzyme is known as enzyme catalysis. Factors affecting enzyme catalysis are:

I.   The rate of enzyme catalyzed reaction is maximum at particular pH, called optimum pH. The pH can increase or decrease the rate of enzyme catalysis.

II. Coenzyme (cofactor) is a non-protein chemical compound that is bound to an enzyme and is required for the biological activity of the enzyme. An inactive enzyme without cofactor is called apoenzyme. Examples: Cu-cytochrome oxidase, Mn-arginose Fe-catalase and Mg-glucose-6-phosphotase.

III. Enzymes catalyzed reactions proceed at high rates and one molecule of an enzyme may transform one million molecules of the substrate per minute.

IV. An enzyme is specific in action. If a compound exists in optically active isomeric forms, an enzyme which acts on one of the isomer will not act on the other isomer.

V. The rate of enzyme catalyzed reaction is maximum at the optimum temperature. In human body, the optimum temperature for enzyme reaction is 37 °C (98.6 F). At high temperatures, all physiological reaction will cease due to the loss of enzymatic activity. Therefore, high body temperature (fever) is unhealthy.

VI. Enzyme activity is greatly influenced by the presence of other substances acting as inhibitors or poisons. The physiological activity of many drugs is related to their action as enzyme inhibitors in body. Example: sulpha drugs inhibit the action of several bacteria and prove effective in curing many diseases. Cyanide acts by blocking the enzyme cytochrome oxidase.

II.15.1 Mechanism and Kinetics of Enzyme Catalysis

Lock-key method

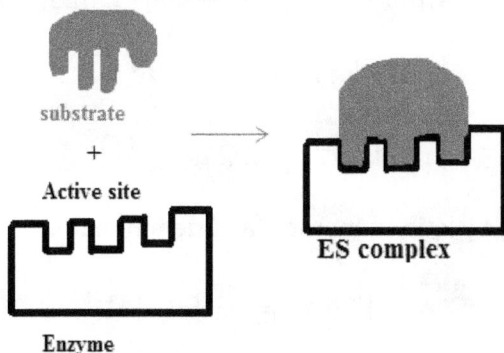

Figure II.14 Lock-key method.

II.16 Steady-State Assumption and Michaelis-Menten Equation

Michaelis-Menten equation describes a curve known as a rectangular hyperbola:

$$E + S \underset{k_{-1}}{\overset{k_1}{\rightleftarrows}} ES \xrightarrow{k_2} E + P$$

where,
E = enzyme, S = substrate, P = product
ES = enzyme-substrate complex
k_1 = rate constant for the forward reaction
k_{-1} = rate constant for the breakdown of the ES to substrate
k_2 = rate constant for the formation of the products

Applying steady-state approximation for intermediate,

Rate formation of [ES] = Rate decomposition of [ES]

$$k_2[S][E] = k_1[ES] + k_{-1}[ES]$$

$$k_2[S][E] - k_1[ES] - k_{-1}[ES] = 0 ...(1)$$

Put $[E]_0 = [E] + [ES]$, where $[E]_0$ is initial concentration of catalyst

$$[E] = [E]_0 - [ES] ...(2)$$

$$k_2[S]([E]_0 - [ES]) - k_1[ES] - k_{-1}[ES] = 0$$

$$k_2[S][E]_0 - k_2[S][ES] - k_1[ES] - k_{-1}[ES] = 0$$

$$k_2[S][E]_0 - [ES]\{k_2[S] + k_1 + k_{-1}\} = 0$$

$$[ES] = \frac{k_2[S][E]_0}{k_2[S] + k_1 + k_{-1}} ...(3)$$

Divide (3) by k_2 in numerator and denominator

$$[ES] = \frac{[S][E]_0}{\frac{[S] + k_1 + k_{-1}}{k_2}} ...(4)$$

Rate of the reaction, $V = K_1[ES]...(5)$

$$V = \frac{k_1[S][E]_0}{\frac{[S]+k_1+k_{-1}}{k_2}} \cdots (6)$$

which is Michaelis-Menten equation.

$$\frac{k_1+k_{-1}}{k_2} = k_m \text{ (Michaelis-Menten constant)}$$

Effect of substrate

Case I
When [S] is small, $k_m >>> [S]$. We can ignore [S] in the denominator of (6) as

$$V = \frac{k_1[S][E]_0}{k_m}$$

$V = $ constant [S]

$V \propto [S]$; first order reaction

Case II
When [S] is large, $[S] >>> k_m$. We can ignore $[k_m]$ in the denominator of (6) as

$$V = \frac{k_1[S][E]_0}{[S]}$$

$V = k_1[E]_0$

$V = $ constant; second order reaction

Also,

$$V_{max} = k_1[E]_0 \cdots (7)$$

Substituting (7) in (6), we obtain

$$V = \frac{V_{max}[S]}{k_m + [S]} \cdots (8)$$

The process is also represented in Figure II.15.

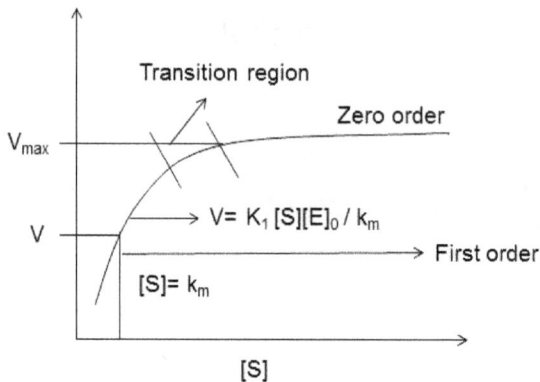

Figure II.15 Effect of substrate on Michaelis-Menten equation.

II.17 Additional Information

Surface tension is the property of a liquid that allows it to resist external forces. Surface tension has the dimension of force per unit length or energy per unit area. Water has the surface energy of 0.72 J/m^2 and a surface tension of 0.72 N/m.

Surface energy: The surface energy was first described by Thomas Yang in 1805. It is the interaction between the forces of cohesion and adhesion which determine whether or not wetting occurs (cohesion: intermolecular attraction between like molecules). The surface energy quantifies the disruption of intermolecular bonds that occurs when a surface is created.

UNIT III

ALLOYS AND PHASE RULE

III.1 Introduction

Metal alloys are a mixture of two or more metals or a mixture of a metal and another element. Some alloys contain only one metal mixed with other non-metals. For example, cast iron is an alloy made of iron metal mixed with carbon non-metal. More precisely, *an alloy is a material comprised of at least two different chemical elements, one of which is a metal.*

➢ The metallic component of an alloy occupying major proportion (90% or more) of the material is called the **main metal, or parent metal, or the base metal.** The other metals or non-metals present in smaller quantities are called alloying **agents**.

➢ Alloy can also be a compound generated by chemical bonding between the components.

➢ Alloy may also be a solid solution: one element dispersed in other element, like salt dissolved in water.

Gold used in a day to day life is an alloy. We refer to how much gold is in the alloy as its karat. Pure gold is the full 24 parts out of 24 karats and 14 karat is 14 parts gold out of 24.

III.2 Significance of Alloying

The common metals such as iron, copper, zinc, lead, tin, aluminium, antimony, magnesium, nickel, manganese, etc., have limited range of properties with respect to the large spectrum of application needs. Hence, these metals have limited use in industrial products in an unalloyed form. Thus, alloys are generated in order to:
➢ enhance the hardness of a metal: an alloy is harder than its components. Pure metals are generally soft. The hardness of a metal can be enhanced by alloying it with another metal or non-metal.

➤ lower the melting point: pure metals have a high melting point. The melting point lowers when the pure metals are alloyed with other metals or non-metals. This makes the metals easily fusible. This property is utilized to make useful alloys called solders.

➤ enhance tensile strength: alloy formation increases the tensile strength of the parent metal.

➤ enhance corrosion resistance: alloys are more resistant to corrosion than pure metals. Metals in pure form are chemically reactive and can be easily corroded by the surrounding atmospheric gases and moisture. Alloying a metal increases the inertness of the metal, thereby enhancing the corrosion resistance.

➤ modify colour: the colour of pure metal can be modified by alloying it with other metals or non-metals containing suitable colour pigments.

➤ provide better castability: one of the most essential requirements of castings is the expansion of the material on solidification. Pure molten metals undergo contraction on solidification. Hence, the metals are needed to be alloyed to obtain good castings because alloys expand on solidification.

III.3 Properties of Alloys

Alloys have unique properties as compared to their individual components like strength, malleability, visual attractiveness, etc. For example, copper and tin are used to make **bronze,** an alloy harder than copper. In addition, alloys have higher hardness, enhanced corrosion resistance, better processasibility and lower melting point than pure metals.

III.4 Structure of Alloys

III.4.1 Substitution Alloys

In substitution alloys, few atoms of the main metal are replaced by the alloying agent. Substitution alloys will form only if the atoms of the base metal and those of the alloying agent are of similar size. Brass is a substitution alloy of copper in which 10-35% of the copper atoms are replaced by zinc atoms. Copper and zinc are close to each other in the periodic table and have similar atomic size.

III.4.2 Interstitial Alloys

Interstitial alloys can form if the atoms of the alloying agent has much smaller size than the base metal. In interstitial alloys, the alloying agents occupy the interstitial sites of the crystalline lattice of the main metal, as shown in Figure III.1. For example, steel is an interstitial alloy of iron and carbon, in which few carbon atoms occupies interstitial sites in the crystalline lattice of iron.

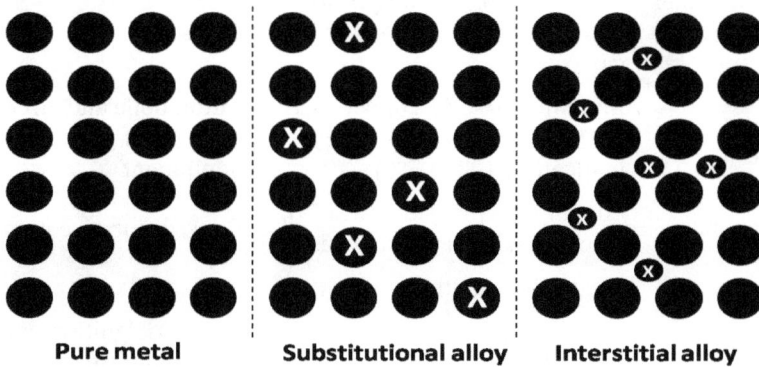

Pure metal   **Substitutional alloy**   **Interstitial alloy**

Figure III.1 Type of alloys.

Table III.1 also lists some important alloys and their diverse applications.

Table III.1 Important alloys and their applications

Alloy	Components	Typical uses
Alnico	Iron (50%+), aluminum (8-12%), nickel (15-25%), cobalt (5-40%), along with other metals such as copper and titanium	Magnets in loudspeakers and pickups in electric guitars
Amalgam	Mercury (45-55%), along with silver, tin, copper and zinc	Dental fillings
Babbitt metal ("white metal")	Tin (90%), antimony (7-15%), copper (4-10%)	Friction-reducing coating in machine bearings
Brass	Copper (65-90%), zinc (10-35%)	Door locks and bolts, brass musical instru-

		ments, central heating pipes
Bronze	Copper (78-95%), tin (5-22%), plus manganese, phosphorus, aluminum or silicon	Decorative statues, musical instruments
Cast iron	Iron (96-98%), carbon (2-4%), along with silicon	Metal structures such as bridges and heavy-duty cookware
Cupro-nickel (copper nickel)	Copper (75%), nickel (25%), including small amounts of manganese	Coins
Duralumin	Aluminum (94%), copper (4.5-5%), magnesium (0.5-1.5%), manganese (0.5-1.5%)	Automobile and aircraft body parts, military equipment
Gunmetal	Copper (80-90%), tin (3-10%), zinc (2-3%) and phosphorus.	Guns, decorative items
Magnox	Magnesium, aluminum	Nuclear reactors
Nichrome	Nickel (80%), chromium (20%)	Firework ignition devices, heating elements in electrical appliances
Nitinol	Nickel (50-55%), titanium (45-50%)	Shape-memory alloy used in medical items, spectacle frames that spring back to shape, and temperature switches
Pewter	Tin (80-99%) with copper, lead, and antimony	Ornaments, used to make tableware before glass became more common
Solder	Old solders contained a mixture of tin (50-70%), lead (30-50%), copper, antimony and other metals. Newer solders have dispensed with lead for health reasons. A typical modern solder has 99.25% tin and 0.75% copper	Connecting electrical components into circuits
Steel (general)	Iron (80-98%), carbon (0.2-2%), including other metals such as	Metal structures, car and airplane parts and many

	chromium, manganese and vanadium	other uses
Steel (stainless)	Iron (50%+), chromium (10-30%), plus smaller amounts of carbon, nickel, manganese, molybdenum and other metals	Jewelry, medical tools, tableware
Stellite	Cobalt (67%), chromium (28%), tungsten (4%), nickel (1%)	Coating for cutting tools such as saw teeth, lathes and chainsaws
Sterling silver	Silver (92.5%), copper (7.5%)	Cutlery, jewelry, medical tools, musical instruments
White gold (18 carat)	Gold (75%), palladium (17%), silver (4%), copper (4%)	Jewelry
Wood's metal	Bismuth (50%), lead (26.7%), tin (13.3%), cadmium (10%)	Solder, melting element in fire sprinkler systems

III.5 Effect of Alloying Elements

Copper

Copper, even in small amounts, provides high corrosion resistance to carbon steel by retarding the rate of rusting at room temperature. High amount of copper in an alloy can cause welding difficulties.

Manganese

Manganese is used up to 1% in all low alloy steels as a deoxidizer and desulphurizer. Manganese increases hardenability and tensile strength of steel, but to a lesser extent than carbon. Manganese readily reacts with sulphur to form manganese sulphide (MnS), which is beneficial for machining as well as surface finish of carbon steel.

Nickel

Nickel enhances ductility or notch toughness of steel. It also increases strength and hardness without sacrificing ductility and toughness. It also provides high corrosion resistance and scaling at elevated temperatures when introduced in suitable quantities in high-chromium (stainless) steels.

Chromium

Chromium, in combination with carbon, is an effective hardening element in the manufacturing of alloys. Chromium increases tensile strength, hardness, toughness, hardenability, resistance to corrosion, resistance to wear and abrasion at elevated temperatures.

Molybdenum

Molybdenum increases strength, hardness, hardenability and toughness, as well as creep resistance at elevated temperatures. It is often used in combination with chromium to improve the strength of the steel at high temperatures. It provides machinability, inhibition to corrosion and red-hardness properties.

Phosphorus

Phosphorus is added to iron as residual element because it reduces ductility and toughness. Small amount of phosphorus is added in steel to increase its strength. It also provides machinability.

Silicon

Silicon is usually alloyed in steel as a deoxidizer and degasifier. Silicon increases tensile and yield strength, hardness, forgeability and magnetic permeability. However, excessive amounts of silicon can reduce the ductility of the alloy. Additional amounts of silicon are sometimes added to welding electrodes to increase the fluid flow of metals.

Aluminum

Aluminum is used in small amounts as a deoxidizer in steel. It may also be added to control the grain size.

Vanadium

Vanadium controls the grain size after heat treatment of steel. It also increases the hardness and resists softening of the steel during tempering treatments. It also increases strength, wear resistance and resistance to shock/impact.

Tungsten

Tungsten is used in steel to enhance strength, wear resistance, hardness and toughness at high temperatures. Tungsten, in combination with carbides, gives higher hardness and exceptional resistance to wear.

Nitrogen

Nitrogen increases the austenitic stability of stainless steel and improves yield strength in such steels, when added in smaller amounts. It also causes brittleness to steel.

Sulphur

Sulphur in small amounts improves machinability, but it is normally an undesirable element in steel as it causes brittleness.

Carbon

Carbon is one of the widely used alloying element because of its effective alloying ability and low cost for increasing the hardness and strength of metals. However, high levels of carbon can deteriorate hardness.

III.6 Nichrome and Stainless Steel

III.6.1 Nichrome

Nichrome is an alloy of Cr (80%) and nickel (20%). It has a high melting point of about 1,400 °C. On heating the blend to red hot temperature, chromium (III) oxide layer forms at the outer surface, which protects inner core materials from further oxidation. Nichrome is widely used in electric heating elements in hair dryers and heat guns due to its low manufacturing cost, strength, ductility, resistance to oxidation, stability at high temperatures and resistance to the flow of electrons.

Chromium trioxide layer formed during heating in air is thermo-dynamically stable and is mostly impervious to oxygen, which protects the heating element from further oxidation, unlike other metals which are oxidized quickly in air upon heating, thus, leading to brittleness and failure.

Physical properties of nichrome

The tensile strength of nichrome is 105,000 PSI (pounds per square inch) and yield strength is around 50,000 PSI. In addition to strength, other useful physical properties of nichrome are presented below:

- Electrical resistivity at room temperature: 1.0×10^{-6} to 1.5×10^{-6} ohm.m
- Thermal conductivity: 11.3 W/m.°C
- Thermal expansion coefficient (20 to 100 °C): 13.4×10^{-6}/°C
- Temperature coefficient of resistivity (25 to 100 °C): 100 ppm/°C
- Specific gravity: 8.4
- Density: 8400 kg/m^3
- Melting point: 1400 °C
- Specific heat: 450 J/kg.°C
- Elastic modulus: 2.2×10^{11} N/m^2

Applications

- Nichrome is used in the explosives and fireworks industry as a <u>bridgewire</u> in electric ignition systems, such as electric matches and model rocket igniters.
- Nichrome wire withstands high temperature, hence, used in ceramics as an internal support structure for clay sculptures to hold their shape.
- Nichrome wire can be used as an alternative to platinum wire for flame testing by colouring the non-luminous part of a flame to detect <u>cations</u> such as sodium, potassium, copper, calcium, etc.
- It is used in motorcycle mufflers and as the heating coils of electronic cigarettes.

III.6.2 Stainless Steel (SS)

An alloy of iron, carbon and chromium is called stainless steel. Steel itself an alloy of iron and carbon. On incorporating chromium to steel, the resulting alloy is termed as stainless steel. Chromium in steel forms a protective layer of chromium oxide known as the 'passive layer', as in the case of nichrome. The chromium oxide layer is very thin, inert and tightly adhered to the metal, which exhibits protection in a wide range of corrosive media.

Other metals are also present in SS depending on its type, e.g., SS 316 contains small amounts of carbon, molybdenum, nickel and manganese. These elements may be added to impart certain properties such as enhanced formability and increased corrosion resistance.

Types of stainless steel

There are five major types of stainless steel.

(a) Austenitic stainless steels contain chromium and nickel with very low carbon content. These are non-magnetic, but can become slightly magnetic when cold worked. Austenitic stainless steels have good corrosion resistance, formability, weldability and excellent mechanical properties over a wide range of temperatures. There are three subtypes in austenitic SS: straight, L and H. (i) Straight type grades are 201, 202, 301, 302, 303, 304, 305, 308, 309, 310, 314, 316, 317, 321, 347, 348 and 384. (ii) L type grades are 304L and 316L; L types have higher corrosion resistance than the straight types. (iii) H types are suitable for use in high temperature environments.

Austenitic stainless steels are used in shafts, valves, bushings, nuts, bolts, aircraft fittings, food processing equipment, chemical equipment, brewing equipment, cryogenic vessels, etc.

(b) Ferritic stainless steels are 400 series. Ferritic stainless steel is chromium containing stainless steel with low carbon content. These are magnetic hardened by cold working. These steels have lower ductility and inferior corrosion resistance than the austenitic grades. However, these offer high resistance to stress corrosion cracking. These steels have weldability limitations which restrict their use to thinner gauges. Some of the commercial grades are 405,

409, 430, 434, 442, 436 and 446. Typical applications of these steel grades are heat exchangers, automotive fasteners, furnace parts, heater parts, etc.

(c) Martensitic stainless steels are 400 series and 500 series. Martensitic stainless steels were the first stainless steel grades commercially developed with relatively high carbon content (0.1-1.2%) and 12-18% chromium. These are magnetic in nature and have higher strength, wear resistance and fatigue resistance than the austenitic and ferritic grades. These steels have moderate corrosion resistance and can be hardened by heat treatment. Some of the commercial grades are 410, 414, 416, 420, 440 and 431. Major applications are machine parts, pump shafts, bolts, bushings, coal chutes, cutlery, hardware, jet engine parts, mining machinery, rifle barrels, screws, valves, aircraft fittings, fire extinguisher inserts, rivets, etc.

(d) Precipitation hardening grade stainless steels are also called PH types. Precipitation hardening stainless steels have been formulated so that these can be supplied in a solution treated condition, (in which these are machineable) and can be hardened, after fabrication, in a single low temperature "ageing" process. Their corrosion resistance is equivalent to that of austenitic grades and strength is generally higher than that of martensitic grades. These steels also retain high strength at elevated temperatures. These are heat treatable and are mainly used in aerospace industry for aerospace structural components. Popular precipitation hardening stainless steel grades include 17-4PH and 15-5PH.

(e) Duplex stainless steels are stainless steels containing relatively high chromium (between 18 and 28%) and moderate amounts of nickel (between 4.5 and 8%). The nickel content is insufficient to generate a fully austenitic structure and the resulting combination of ferritic and austenitic structures is called duplex. Most duplex steels contain molybdenum in a range of 2.5-4%. These provide higher corrosion resistance and resistant to stress corrosion cracking than the austenitic stainless steels.

Popular UNS duplex grades include S32101, S32304, S32003, S31803, S32205, S32760, S32750, S32550, S32707 and S33207. Typical applications of the duplex stainless steel grades include water treatment plants, petrochemical plants as well as heat exchanger components.

III.7 Heat Treatment of Steel

The process of heating and cooling of solid steel articles under controlled conditions, thereby developing certain physical properties, without altering its chemical composition is known as heat treatment. It causes the refinement of grain structure, removal of the entrapped gases and removal of internal stresses. The various heat-treatment processes are as follows:

III.7.1 Annealing

It means **softening.** The process of heating the metal to a certain high temperature, followed by very slow cooling in a planned manner in a furnace is called **annealing**. Annealing process is classified into two categories as follows:

(i) Low temperature annealing: In this process, steel is heated to a temperature below the lower critical point, followed by slow cooling. It improves machinability, ductility and shock resistance, but reduces hardness.

(ii) High temperature annealing: In this process, steel is heated to a temperature about 30-50 °C above the higher critical temperature, followed by slow cooling. It increases ductility, toughness and machinability.

III.7.2 Hardening or Quenching

It is a process of heating steel beyond its critical temperature and then suddenly cooling it using oil or sea water. Medium and high carbon steels can be hardened by rapid cooling, but low-carbon steels cannot be hardened. This process produces hardness and high wear resistance, but it also leads to higher brittleness.

III.7.3 Case Hardening

In this process, the diffusion of an alloying element (carbon or nitrogen) into the surface of a monolithic metal is carried out. The resulting steel has high wear resistant carbon rich surface. This process is adopted for low-carbon steels, which cannot be hardened by quenching process.

The process of case-hardening is carried out in two stages:

***Carburizing*:** The mild steel is heated to 900-950 °C in the presence of charcoal in a cast iron box and kept at these conditions for sufficient time. Subsequently, it is cooled slowly within the box. During this process, carbon diffuses into the surface of steel. This process is called **carburizing**.

***Hardening*:** The carburized article is re-heated to about 900 °C and subsequently quenched in oil so that the brittleness is reduced and the steel becomes tough and soft. The article is then re-heated to about 700 °C and quenched in water.

III.7.4 Flame Hardening

This process involves heating a metal with a high-temperature flame of oxy-acetylene followed by quenching. It is used on medium carbon, mild or alloy steels or cast iron to produce a hard and wear-resistant surface.

III.7.5 Gas Carburizing

It is the process by which carbon is diffused into the surface layer to improve wear and fatigue resistance. The metal is subjected to coal-gas at high temperature, usually above 925 °C, which causes the infusion of carbon into the outer layer.

III.7.6 Tempering

Tempering is used to increase the toughness of iron alloys, particularly steel. It is a process of heating the already hardened steel to a temperature lower than its own hardening temperature and allowing it to cool slowly. Untempered steel is very hard, but is too brittle for most applications. Tempering is commonly performed after hardening to reduce excess hardness and brittleness. The hardened steel is re-heated to the temperature range 400 to 600 °C, thus, resulting in effective ductility and toughness.

III.7.7 Normalizing

The main objective of the normalizing heat treatment is to enhance the mechanical properties of the material by refining the micro-structure. It involves heating steel to a temperature above its higher critical temperature and allowing it to cool gradually in air. It takes much lesser time than the annealing process. It provides homogene-ity to the steel structure, removal of internal stresses, refining of grains and enhancement of toughness. The normalizing heat treat-ment balances the structural irregularities and makes the material soft for further working.

III.7.8 Cyaniding

In this process, an iron-base alloy is heated in contact with a cyanide salt, such that the surface absorbs carbon and nitrogen. During the process, the metal is immersed into a molten KCN or NaCN solution at a temperature of about 870 °C and subsequently quenched in oil or water.

Cyaniding is followed by tempering to obtain a desired combina-tion of hardness and toughness.

III.7.9 Nitriding

Gas nitriding is a surface hardening process, where nitrogen is add-ed to the surface of steel parts using dissociated ammonia as the source, at a temperature of about 550 °C.

Overall, gas nitriding process develops a hard texture in a com-ponent at relatively low temperature, without the need for quench-ing.

III.8 Phases, Components and Phase Rule

Phase rule is a tool to understand the quantitative dependents of individual components of a heterogeneous system in equilibrium. J. Willard Gibbs, a physical chemist, introduced the phase rule theoret-ically in 1876.

III.8.1 Phase

A homogeneous physically distinct and mechanically separable part of the system which is separated from other part of the system by a definite boundary.

Examples of single phase systems:
- Mixture of gases constitute one phase
- Miscible liquids such as alcohol and water constitute one phase
- Solution of solid substance in liquid constitutes one phase (salt in water)

In all the above examples it is not possible to physically separate the constituents from each other.

Examples for two phase systems:
- Immiscible liquids such as CCl_4 and water
- Rhombic sulphur and monoclinic Sulphur
- Emulsion of oil in water
- Water and water vapour

In all the above examples we can visualize and mechanically separate the constituents from each other.

Examples for three phase systems:

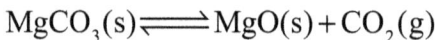

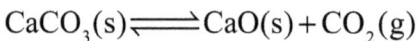

$$ice \rightleftharpoons water \rightleftharpoons water \text{ vapour}$$

$$MgCO_3(s) \rightleftharpoons MgO(s) + CO_2(g)$$

$$CaCO_3(s) \rightleftharpoons CaO(s) + CO_2(g)$$

In the above systems 2 and 3, two solid constituents are separable from each other, i.e., MgO (or CaO) is separable from $MgCO_3$ (or $CaCO_3$), and so forms two distinct solid phases. Thus, the number of phases for the above systems are three (two solid phases + one gas phase).

III.8.2 Components

In a system at equilibrium, the composition of each phase can be expressed in the form of a chemical formula of the independent species (constituents) present in the system. The smallest number of such independent species are called components of the system. In other words, the minimum number of constituents required to define the composition or concentration of the all the species in the system is termed as components.

For example,

$$ice \rightleftharpoons water \rightleftharpoons water \text{ vapour}$$

All of the three phases can be expressed as H_2O, so the number of components is 1.

Consider a chemically reactive system at equilibrium

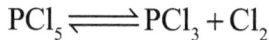

$$PCl_5 \rightleftharpoons PCl_3 + Cl_2$$

The above system consists of three constituents, however, if we know the concentration of PCl_5 and PCl_3, the concentration of Cl_2 is automatically fixed. Thus, the composition of this system can be expressed using PCl_5 and PCl_3 alone and hence the number of components is 2.

Similarly,

$$CaCO_3 \rightleftharpoons CaO + CO_2$$

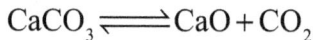

Here, the composition of above three phases can be expressed as

$$CaCO_3 \rightleftharpoons CaCO_3 + 0CaO$$
$$CaO \rightleftharpoons 0CaCO_3 + CaO$$
$$CO_2 \rightleftharpoons CaCO_3 - CaO$$

The composition of the whole system can be expressed in terms of $CaCO_3$ and CaO alone. This means that the concentration of CO_2 is automatically known. Thus, the number of components is 2.

The number of components is expressed by the following equation

$$C = N - E$$

where,
C is number of components
N is number of chemical species
E is the number of independent equations required to define the system completely

III.8.3 Degrees of Freedom

Degrees of freedom (F) is defined as the number of independent variables such as temperature, pressure and concentration, which must be specified in order to define a system completely.

Consider water in equilibrium with its vapour

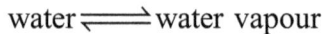

$$water \rightleftharpoons water\ vapour$$

At a given temperature, the equilibrium vapour pressure of water can have only one fixed value, i.e., if the temperature is 100 °C, the vapour pressure is 1 atm. Thus, if temperature is specified, the pressure becomes known automatically and vice versa. Since one variable (temperature or pressure) is enough to define this system at equilibrium, the number of degrees of freedom for this system is 1.

Consider another 3 phase system of water

$$ice \rightleftharpoons water \rightleftharpoons water\ vapour$$

Here, the equilibrium among the three phases can exist only at particular temperature and pressure (4.58 mm and 0.0075 °C). If we change any one variable, the the equilibrium among the 3 phases will not exist. None of the three variable can be specified to define this system and, hence, the degrees of freedom is zero. This is an invariant system.

Degrees of freedom can also be calculated using phase rule.

III.8.4 Phase Rule

The equilibrium condition of a heterogeneous system is influenced by temperature, pressure and composition. Under such conditions, the degrees of freedom (F) of a system is related to number of components (C) and number of phases (P) by the following equation, provided the equilibrium between the phases is not influenced by electrical, magnetic or gravitational forces.

$$F = C - P + 2$$

$$ice \rightleftharpoons water \rightleftharpoons water \ vapour$$

Number of components, C = 1 (H_2O)
Number of phases, P = 3 (ice (s), water (l), water vapour (g))
Degrees of freedom, F = 1 − 3 + 2 = 0

Advantages of phase rule

- Predicting the behaviour of the system is easy
- It is applicable to chemical as well as physical equilibria
- Data about molecular structure is not required
- It does not depend on the nature or total quantity of the system
- It provides information about phase changes in terms of temperature, pressure and composition.

Limitations of phase rule

- It can be applied only for systems which are in equilibrium
- It considers only three variables namely temperature, pressure and composition. It fails to consider electrical, magnetic and gravitational forces and surface actions.

Derivation of phase rule

Degree of freedom = total number of independent variables − total number of independent equations

Total number of variables: System at equilibrium has temperature, pressure and composition as variables. At equilibrium, each phase has same temperature, thus, there is one temperature variable. At equilibrium, each phase has same pressure, there there is one pressure variable.

Consider a single phase system and two components A & B. If concentration of A is 0.6, then the concentration of B should be 1-0.6 = 0.4. If we know the concentration of one, other will be automatically known. Thus, if we have C components, there must be C-1 composition variables. For P phases, the total number of compositions are P(C-1).

Thus, total number of variables =1 (for T) + 1 (for P) + P(C-1)

Total number of relations (equations)
Consider,

$$PCl_5(s) \rightleftharpoons PCl_3(g) + Cl_2(g)$$

To express the composition of each component, there must be P-1 separate equations. Hence, for C components, the number of such equations is C(P-1).

Substituting the number of variables and number of relations in the earlier equation:

Degrees of freedom = [1 (for T) + 1 (for P) + P(C-1)] - C(P-1)
F = 2+PC-P-CP+C
Thus, F = C-P+2

III.9 Phase Diagram

A phase diagram is a graphical representation of systems in chemical equilibrium with respect to variables such as temperature, pressure and composition. From phase diagram, we can predict the temperature and pressure for a particular composition to exist in equilibrium.

III.9.1 One Component System

Well known example for a one component system is water system.

$$ice \rightleftharpoons water \rightleftharpoons water \ vapour$$

In this system, the number of components is one and number of phases is three.
C = 1 (H_2O), P = 3 (ice (s), water (l), water vapour (g))
Thus, F = 1-3+2 = 0

The phase diagram of water is given in Figure III.2.
Areas:

        BOC: H_2O in solid phase (ice)
        AOC: H_2O in liquid phase (water)
        AOB: H_2O in gas phase (vapour)

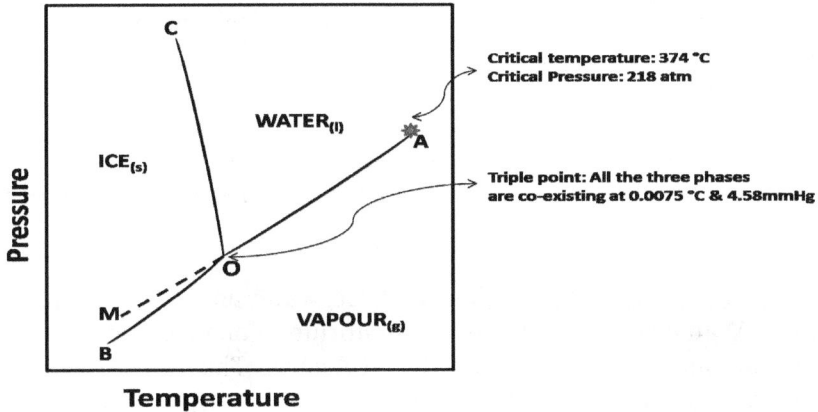

Figure III.2 Phase diagram of water.

Curves:
Curve AO is sublimation curve of ice and water vapour co-existing in equilibrium. The lower limit of this curve is absolute zero.

$$liquid \rightleftharpoons vapour \ (water \rightleftharpoons water \ vapour)$$

For this system, P = 2 and C = 1. Thus, on this curve, two phases co-exist in equilibrium, so F = 1 – 2 + 2 = 1.

Curve OB is vapour pressure curve or vapourization curve of water. Along this curve, liquid water and water vapour co-exist in equilibrium.

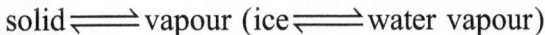

$$solid \rightleftharpoons vapour \ (ice \rightleftharpoons water \ vapour)$$

For this system, P = 2 and C = 1. Thus, on this curve, two phases co-exist in equilibrium, so F = 1 – 2+2 = 1.

Curve OC is called melting point curve. The ice and water are in equilibrium along this line. Thus, curve OC is slightly inclined towards pressure axis indicating that melting point of ice decreases with increase of pressure.

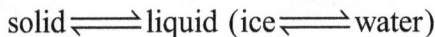

$$solid \rightleftharpoons liquid \ (ice \rightleftharpoons water)$$

For this system, P = 2 and C = 1. Thus, on this curve, two phases co-exist in equilibrium, so F = 1 – 2 + 2 = 1.

Overall, three equilibria can exist in water system. On each curve, two phases co-exist in equilibrium, so F = 1 – 2 + 2 = 1. To define any point along a line, it is enough to specify any one variable (temperature or pressure).

Point "O" (triple point): at this point, ice, water and water vapour co-exist. Water and vapour are in equilibrium along OB curve and ice and vapour are in equilibrium along OA curve. Thus, all three phases of water system co-exist at point O, called as triple point. At this point, temperature and pressure are 0.0075 °C and 4.58 mmHg.

$$ice \rightleftharpoons water \rightleftharpoons water \ vapour$$

For this case, F = 1 – 3 + 2 = 0. As the degree of freedom is zero, this is non-variant point.

Meta stable equilibrium curve (OM): water can be cooled up to -9 °C without formation of ice. This is called supercooled water. The vapour pressure curve OB can continue below the point O, as shown as dotted line. Along the curve OM, supercooled water and vapour are in equilibrium, which is called metastable equilibrium.

III.9.2 Two Component Alloy Systems

We know that,

Degrees of freedom = components – phases + variables (temperature & pressure) or

$$F = C - P + 2$$

Condensed phase rule: A solid equilibrium of an alloy has practically no gaseous phase and the effect of pressure is negligible. Therefore, the experiments are conducted under atmospheric pressure. Thus, the gas phase is ignored. Since the pressure is kept constant, thus, the phase rule becomes

$$F = C - P + 1$$

This equation is called **condensed phase rule** or **reduced phase rule.** Therefore, the equilibrium in two component alloy are described with only two variables, namely, temperature and concentrations.

The possible equilibrium between the phases of two component alloy system is
- Solid – liquid equilibria
- Liquid – liquid
- Liquid – liquid + solid

Classification of alloy system: Binary composition of metals can form homogeneous mixture on melting, depending upon their miscibility, solubility and reactivity. The mixture can be classified into following types:
- Simple eutectic formation
- Formation of compound with congruent melting point
- Formation of compound incongruent melting point
- Formation of solid solution

Simple eutectic system is a binary system consisting of two substances, which are miscible in all proportions in liquid phase, but which do not react chemically. Consider two metals A and B, and different proportions of AB is prepared as follows:

A	100	90	80	70	60	50	40	30	20	10	0
B	0	10	20	30	40	50	60	70	80	90	100
FP of AB	270	250	230	220	200	225	248	280	321	345	362

The freezing point (FP) of A is 270 and freezing point of B is 362. As we know that the addition of a substance (impurity) to a pure substance results in lowering of freezing point of the pure substance. (freezing point: temperature at which liquid to solid transformation takes place). Similarly, the freezing point of A 270 is decreased when B is added in to it. Among the different proportions of AB, the mixture having the lowest freezing point is known as the eutectic mixture. This minimum freezing point corresponding to eutectic mixture is termed as "eutectic point".

In the above example, The mixture A:B = 60:40 wt% exhibits lowest freezing point of 200 °C, thus, this is the eutectic mixture of the AB alloy system.

In a eutectic system,

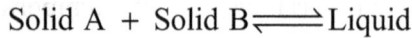

$$\text{Solid A} + \text{Solid B} \rightleftharpoons \text{Liquid}$$

III.10 Construction of Phase Diagram by Thermal Analysis (or) Cooling Curve

III.10.1 Thermal Analysis

Thermal analysis is the study of the cooling curves of various compositions of a system during solidification. The data obtained from thermal analysis along with recorded curves are called as thermograms. These thermograms are characteristic of a particular system composed of either single or multi-component materials. For any mixture of a definite composition, it is possible to find out freezing and eutectic points from the cooling curves. Thermograms indicate the system in terms of temperature dependencies of its thermodynamic properties (Figure III.3).

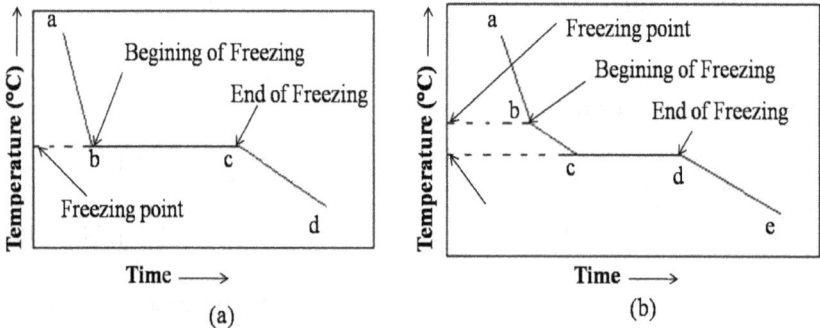

Figure III.3 Cooling curves.

Cooling curve Figure III.3 (a): A metallic substance in molten state is cooled slowly and the temperature is noted at different time intervals. Subsequently, the change in temperature is plotted against time, as shown in Figure III.3 (a). In this diagram, 'ab' denotes the rate of cooling of molten liquid and the liquid starts solidifying at the freezing point 'b'. Now, the temperature remains constant until the liquid melt is completely solidified. Solidification completes at the point 'c'. The horizontal line 'bc' represents the equilibrium between the solid and liquid melt. After the point 'c', the temperature of the solid begins to decrease along the curve 'cd'.

Cooling curve Figure III.3 (b): When a molten liquid containing two components (say A and B) is cooled slowly, the observed cooling curve is different from cooling curve of pure metal, as seen in Figure III.3 (b). Initially, the rate of cooling is continuous, similar to pure substance. As it reaches the point 'b', one of the substances (either A or B) begins to solidify out of the melt, which is indicated by a break and the rate of cooling is different. On further cooling at the break point 'c', the second compound also begins to solidify. Now, the temperature remains constant until the liquid melt is completely solidified, which forms the eutectic mixture (line 'cd'). After the break point 'd', cooling of solid mass begins. The temperature of horizontal line 'cd' provides the eutectic temperature. The temperature measurements are performed with a sensitive thermometer and the arrest points are determined with good precision.

A number of mixtures of A and B are taken with different composition. Each mixture is heated to the molten state and the cooling curves are drawn separately for each mixture. From the cooling curves of various compositions, the main phase diagram can be drawn by taking the composition on X-axis and temperature on Y-axis. Any point on this line indicates the appearance of the solid phase from the liquid. The area above this curve is only liquid phase.

III.10.2 Uses of Cooling Curves

- Cooling curves are used to find the % purity of the compounds.
- These are used to find the melting point of the compounds.
- Thermal analysis is useful in derivation of phase diagram of any two component system.
- These are used to find the composition of the alloys.
- The cooling curves are used to analyze the behaviour of the compounds.

III.10.3 Lead-Silver Alloy System

It is a two component system with four possible phases – solid Ag, solid Pb, solution of Ag + Pb and its vapour. The two metals are completely miscible with each other in liquid state and do not form any chemical compound. There is almost no effect of pressure on equilibrium, thus, the temperature and composition are considered to construct the phase diagram at constant atmospheric pressure. The phase diagram of lead-silver system is shown in Figure III.4.

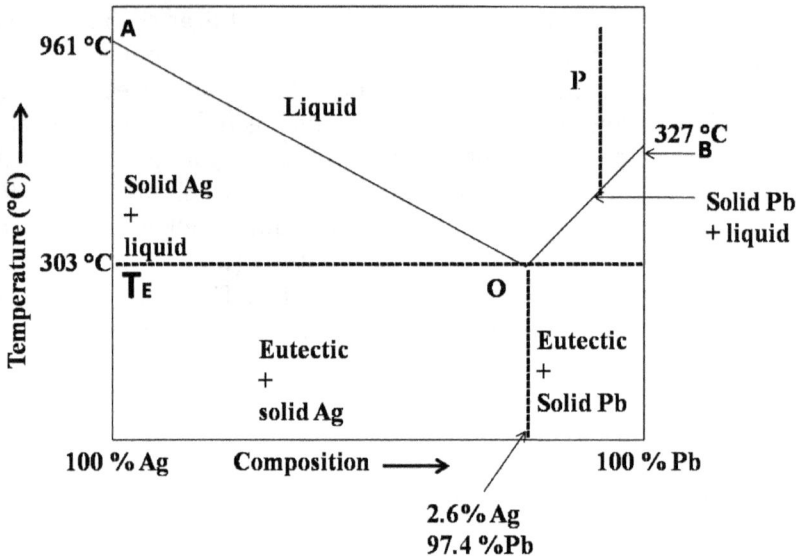

Figure III.4 Phase diagram of lead-silver system.

The point A is the freezing point of pure silver, point B is freezing point of pure lead. The area above AOB consists of liquid phase only (Pb, Ag melt). The components Ag and Pb exist as solution. Thus, C = 2 and P = 1. Hence, F = C − P + 1 = 2 − 1 + 1 = 2, the system is bivariant. The system will exist when the temperature T > 303 °C ,Pb < 97.4 % and Ag > 2.6 %. The area below XOY is completely solid phase, whereas the area AOY consists of solid Ag and melt.

Curve AO: It is the freezing point curve of Ag (961 °C). The curve shows decrease in freezing point/melting point of Ag due to the addition of Pb to Ag. Along this curve, solid Ag is in equilibrium with solution of Pb in Ag.

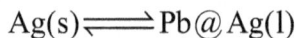

$$Ag(s) \rightleftharpoons Pb@Ag(l)$$

Curve BO: The BO curve is the freezing point curve of Pb (327 °C).

It shows decrease in freezing point/melting point of Pb due to the addition of Ag to Pb. Along this curve, solid Pb is in equilibrium with solution of Ag in Pb.

$$Pb(s) \rightleftharpoons Ag @ Pb(l)$$

Here, C = 2 and P = 2, thus, the reduced phase rule is F = C – P + 1 = 2 – 2 +1 = 1. Hence the system is **univariant**. The point O (303 °C) represents a fixed composition of 97.4% Pb and 2.6% Ag, and is called **eutectic composition**. On cooling, the whole mass crystallizes out as such.

Eutectic point 'O': The curves AO and BO meet at the point O, called as eutectic point. Here, solid Ag, solid Pb and solution of Ag & Pb are in equilibrium. Thus, C = 2 and P = 3. Hence, the reduced phase rule is F = C – P + 1 = 2 – 3 + 1 = 0 and the system is **invariant**. The point O (303 °C) represents a fixed composition of 97.4% Pb and 2.6% Ag, and is called **eutectic composition**.

Area above AOB: The components Ag and Pb exist as solution. Thus, C = 2 and P = 1. Hence, F= C – P + 1 = 2 – 1 + 1 = 2, and the system is bivariant. The system will exist when the temperature T > 303°C, Pb < 97.4% and Ag > 2.6%.

Pattinson's process: The phase diagram of lead-silver is useful in the extraction of silver from the argentiferous lead ore which has a very small percentage of silver (0.1%). The alloy is heated above 327 °C and is subsequently allowed to cool. The melted alloy reaches 'O' on the curve BO, solid Pb separates out and the solution has more Ag. On further cooling, more of Pb is separated till the eutectic point is reached. At 'O', an alloy containing 2.6% Ag and 97.5% Pb is obtained. This process is known as Pattinson's process.

UNIT IV

FUELS AND COMBUSTION

IV.1 Introduction

A fuel is a carbon rich combustible material, which on burning produces heat and light energy. The most important sources of fuel are coal and gasoline. These are stored fuels under the earth's crust. The main characteristics of fuels include

- ❖ High calorific value
- ❖ It must be in safe for storage and transport
- ❖ Moderate ignition temperature
- ❖ Readily available and cheap
- ❖ The products of combustion must not be harmful to environment and human health

Fuels are classified into three types based on the three states of materials as

- ❖ Solid fuels
- ❖ Liquid fuels
- ❖ Gaseous fuels

IV.2 Solid Fuels

Solid fuels are mainly classified into two types, i.e. natural fuels, such as wood, coal, etc., and manufactured fuels, such as charcoal, coke, briquettes, etc.

Advantages of solid fuels

- ❖ The solid fuels are easy to transport
- ❖ These are convenient to store without substantial risk of explosion/fire
- ❖ The cost of production is low
- ❖ They possess moderate ignition temperature

Disadvantages of solid fuels

❖ Their ash content is high
❖ The large proportion of heat is wasted
❖ The solid fuels burn with clinker formation
❖ Their combustion operation cannot be controlled easily
❖ Their handling cost is high

IV.2.1 Coal

Coal is a highly carbonaceous solid fuel formed under the earth crust as a result of modification of decomposed vegetable matter under high pressure and temperature.

Coal is also defined as a metasedimentary rock, which means that it is formed by sedimentary process and metamorphosed. Coal forms when dead plant mass is converted into **peat,** which is then transformed into **lignite**, followed by its conversion to **sub-bituminous** coal, and finally sub-bituminous coal gets converted to **anthracite**. This biological and geological process of conversion of peat to anthracite over a period of time is known as metaphmorphism.

Peat ➞ Lignite ➞ sub-bituminous ➞ Anthracite

Metamorphism
The word "Metamorphism" comes from the Greek: meta = after, morph = form, so metamorphism means the after form. In geology, a rock formed at certain temperature and pressure may undergo a change in its morphology or assemblage and texture when subjected to pressure and temperature ranges higher than prevalent during the formation of the rock.

Lignite, known as brown coal, is formed from naturally compressed peat. It has low heat content and a carbon content of 70%.

Bituminous coal (black coal), is formed from lignite, a soft coal. It is of higher quality than lignite and inferior to anthracite.

Anthracite is formed from bituminous coal. It has highest carbon content of 92-98% and has highest calorific value.

IV.2.2 Analysis of Coal

To assess the commercial value of coal, certain analyses on its burning properties are essential. Two commonly used tests are (1) proximate analysis and (2) ultimate analysis of coal.

Calorific value of coal is defined as the quantity of heat produced by burning a unit weight of coal in a calorimeter.

Proximate analysis

Proximate analysis provides information about heating and burning properties of coal. The test provides the composition of coal in respect of the following:

❖ humidity content (moisture)
❖ explosive matter (volatile matter)
❖ dust content (ash)
❖ permanent carbon in coal (fixed carbon)

Humidity content: 1 g of powdered, air-dried coal sample is taken in a silica crucible and is heated at 100 -105 °C in an electric hot air oven for 1 hour. Afterwards, the crucible is cooled in an desiccator and weighed. The % of humidity is calculated from the loss in weight of the sample as

$$\% \ of \ humidity \ in \ coal = \frac{loss \ of \ weight \ of \ the \ coal}{weight \ of \ air \ dried \ coal} \times 100$$

Explosive matter: after the analysis of humidity content, the crucible with remaining coal sample is enclosed with a lid and is heated at 950 ± 20 °C for 8 minutes in a muffle furnace. The amount of explosive matter in the sample is calculated from the loss in weight of the sample as

$$\% \ of \ explosive \ matter \ in \ coal = \frac{loss \ of \ weight \ of \ the \ coal}{weight \ of \ air \ dried \ coal} \times 100$$

Dust content (ash): after the analysis of explosive matter, the crucible containing residual coal sample is heated in open at 700 ± 50 °C for 1 hour in a hot furnace. The % of dust content is calculated from the loss in weight of the sample as

$$\% \ of \ dust \ in \ coal = \frac{\text{loss of weight of the coal}}{\text{weight of air dried coal}} \times 100$$

Permanent carbon in coal: it is determined by subtracting the sum total of humidity, volatile and dust contents from 100.

Ultimate analysis

It is the analysis involving the determination of chemical constituents such as:

* ❖ Carbon and hydrogen content
* ❖ Nitrogen content
* ❖ Sulphur content
* ❖ Oxygen content
* ❖ Ash content

Carbon and hydrogen content: a known quantity of coal is burnt in the presence of oxygen in a combustion chamber. The carbon and hydrogen present in the coal sample are converted into CO_2 and H_2O respectively, as given below:

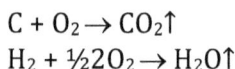

$$C + O_2 \rightarrow CO_2\uparrow$$
$$H_2 + \tfrac{1}{2}2O_2 \rightarrow H_2O\uparrow$$

The CO_2 and H_2O vapours are immersed correspondingly in KOH and anhydrous $CaCl_2$ tube of known weights. The increase in weight of KOH tube is due to the absorption of CO_2, while the increase in weight of $CaCl_2$ tube is due to the absorption of H_2O. From the weights of CO_2 and H_2O formed (as given in above equations), the % of carbon and hydrogen content in coal can be calculated.

Nitrogen content: the determination of nitrogen content is carried out by Kjeldahl's method. A known amount of compressed coal is heated with concentrated H_2SO_4 acid in the presence of K_2SO_4 in a flask. Nitrogen in the coal is converted into ammonium sulphate as

$$2N_2 + 3H_2 + H_2SO_4 \rightarrow (NH_4)_2SO_4$$

The clear solution is heated with excess of NaOH and ammonia is distilled over and is absorbed in a known volume of N/10 HCl.

$$(NH_4)_2SO_4 \rightarrow 2NH_3 + NaSO_4 + 2H_2O$$
$$NH_3 + HCl \rightarrow NH_4Cl$$

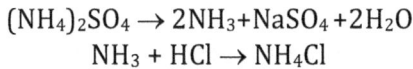

The quantity of unused N/10 HCl is determined by titrating it against standard N/10 NaOH. Thus, the quantity of acid neutralized by liberated ammonia from coal is determined, which is then used to calculate the percentage of nitrogen.

Sulphur content: sulphur in coal is converted into sulphate and is extracted with water. The water extract is treated with $BaCl_2$ solution, which converts the sulphates into $BaSO_4$. The $BaSO_4$ precipitate is filtered, dried and weighed. From the amount of $BaSO_4$ obtained, the sulphur content in the coal is calculated.

Ash content: determination of ash substance is carried out as in proximate analysis

Oxygen content: the percentage of oxygen content in coal is calculated as
% of oxygen in coal=100 - % of (C+H+N+S+ash)

Significance of ultimate analysis

a) Carbon and hydrogen content
 ❖ Higher the % of carbon and hydrogen, better is the quality of coal and higher energy value
 ❖ The % of carbon is useful in the classification of coal
 ❖ Higher percentage of carbon in coal requires smaller size burning chambers

b) Nitrogen content
 ❖ Nitrogen does not have any calorific value and its presence in coal is undesirable
 ❖ High-quality coal should be free from nitrogen content

c) Sulphur content
 ❖ The burning products of sulphur are damaging and corrosive, thus, contributing to environmental pollution
 ❖ The coal containing sulphur is not recommended for the preparation of metallurgical coke because sulphur can degrade the properties of the metal

d) Oxygen content
  ❖ Oxygen content decreases the calorific importance of the fuel.
  ❖ As the oxygen content increases, the humidity content also increases and the calorific value of the fuel decreases.

IV.2.3 Carbonization

The process of removing moisture content and volatile matter from coal by heating in the absence of air, in order to change it into coke, is called carbonization. Carbonization of coal is also known as coking of coal.

Caking coals and coking coals

The coal to coke transformation takes place when the coal is heated. At the fusing temperature, the coal becomes flexible and plastic, thereby fusing to generate a coherent mass. Such type of coals are called caking coals. If the residue (coke) obtained after carbonization is hard, strong, porous and suitable for metallurgical application, then the original coal is called as coking coal.

Types of carbonization

Based on the applied temperature, carbonization is classified into two types, low temperature carbonization (LTC) and high temperature carbonization (HTC).

In LTC, carbonization is carried out at 500-700 °C. Yield of the coke is in the range 75-85%. The resulting coke is soft and has low mechanical strength and low calorific value. The formed coke is observed to burn in a smokeless manner.

In HTC, carbonization is carried out at 900-1300 °C. Yield of the coke is in the range 65-75%. The formed coke is hard and has high mechanical strength and high calorific value. In addition, it is observed to burn in a smoky manner.

Metallurgical coke

The coal (preferably bituminous) is heated in vacuum. The volatile matter evaporates and the coal becomes porous, hard and strong, and is termed as metallurgical coke.

Properties of good metallurgical coke

- ❖ Purity: The moisture, ash, sulphur and phosphorus sub-stances in metallurgical coke must be low
- ❖ Porosity: Coke must be highly porous so that oxygen will have intimate contact with carbon and burning will be abso-lute and homogeneous
- ❖ Strength: The coke should have very high strength to sustain high pressure of the overlying material in the furnace.
- ❖ Calorific value: The calorific value of coke must be very high
- ❖ Combustion: The coke must flame easily
- ❖ Reactivity: The reactivity of the coke must be small because low reactive coke materials produce high temperature on combustion
- ❖ Cost: It must be cheap and abundantly available

Manufacture of metallurgical coke

Otto Hoffman's byproduct oven

To enhance the thermal efficiency of carbonization process and to recover the valuable products such as coal gas, ammonia, benzol oil, tar, etc., Otto Hoffman developed byproduct coke oven, as shown in Figure IV.1.

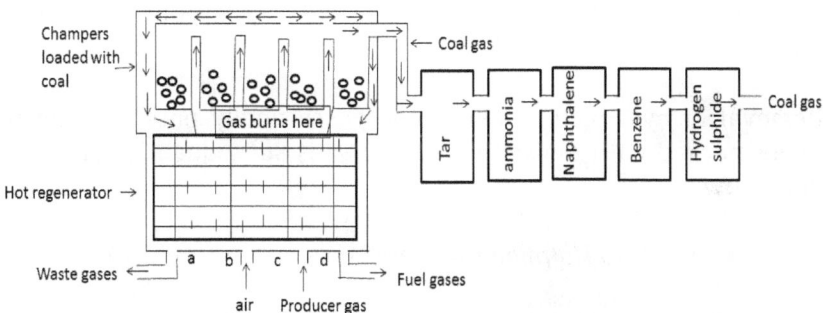

Figure IV.1 Otto Hoffman byproduct coke oven.

The oven consists of a series of silica chambers connected to-gether as shown in Figure IV.1. Coal is loaded into the silica chamber and heated to 1200 °C by burning the producer gas mixture along the sides of the chambers.

The air and producer gas are preheated by sending through 'b' and 'c' hot regenerators. Hot flue gases produced during combustion are passed through 'a' and 'd' regenerators until the temperature reaches 1000 °C, while 'a' and 'd' regenerators are being heated by hot flue gases. After completing this process, the coke is removed and quenched with cold water. *Otto Hoffman's* method produces coke with 70% yield within 12-20 hours of carbonization period. The byproducts such as tar, ammonia, H_2S and benzene are recovered from coal gas as follows.

Recovery of byproducts

Tar: the gas from the coke ovens is passed to recovery towers in which ammonia liquor trickles from the top. Tar and dust dissolve in ammonia and are collected in a tank.

Ammonia: the gases subsequently enter into 2nd tower where water is sprayed from top and ammonia is converted into NH_4OH. Dilute H_2SO_4 may also be employed in the process and ammonium sulphate is recovered.

Naphthalene: the gases are then passed to 3rd tower, where cooled water is sprayed to condense naphthalene.

Benzene and its derivatives: the gases are passed through 4th tower, where petroleum is sprayed. Here, benzene vapors get condensed to liquid.

Removal of hydrogen sulphide: the gases finally pass through 5th tower packed with moist Fe_2O_3 and H_2S is subsequently recovered in the tower.

Advantages of Otto Hoffman's process

❖ The important byproducts such as coal gas, naphthalene, ammonia and H_2S are recovered and appropriately utilized, thus, reducing wastage
❖ Heating is performed in four separate chambers by producer gas and preheated gas mixture
❖ The carbonization time in the process is low in the range of 12-20 hours

IV.3 Liquid Fuels

IV.3.1 Petroleum

Petroleum or crude oil is a naturally available liquid fuel under the earth crust. It is a shady brown or black coloured gelatinous oil formed deep in the earth crust. The oil is regularly floating over a salt water solution and natural gas is present above the oil. Oil is a mixture of paraffinic, olefinic and aromatic hydrocarbons in addition to small amounts of N, O and S.

Types of petroleum

Petroleum is classified into three types:

- ❖ Paraffinic petroleum oil: It is mainly composed of saturated hydrocarbons from CH_4 to $C_{35}H_{72}$ with a smaller amount of naphthenes and aromatics.
- ❖ Naphthenic or asphaltic petroleum oil: It contains cycloparaffins or naphthenes with a small amount of paraffins and aromatics.
- ❖ Mixed base petroleum oil: It contains equal quantity of paraffinic and asphaltic hydrocarbons.

IV.3.2 Distillation of Petroleum or Crude Oil

The crude oil obtained from the earth is a mixture of oil, water and unwanted impurities. After the elimination of water and other impurities, the crude oil is subjected to fractional distillation. During fractional distillation, the crude oil is fractionated into several fractions.

The process of removing impurities and separating the crude oil into several fractions having different boiling points is known as refining of petroleum.

Step 1. Separation of water (Cottrell's process): The crude oil is extracted from oil well as an emulsion of oil and salt water. As a first step, the crude oil is allowed to flow between charged electrodes, so that the small colloidal water droplets merge to form large drops. Subsequently, these large water drops are separated out from the oil.

Step 2. Removal of harmful sulphur compounds: Sulphur compounds are removed by treating the crude oil with copper oxide. Copper sulphide, produced as a solid, is separated out by filtration.

Step 3. Fractional distillation: The purified crude oil is then heated to about 400 °C in an iron furnace, where the oil gets vaporized. The hot vapours are passed into the bottom of a "fractionating column". The fractionating column is a tall cylindrical tower containing horizontal stainless steel trays at short distances. Each tray is provided with small chimney covered with a loose cap. The vapours of the oil entering the fractionating column become cooler and get condensed at different trays. The fractions having higher boiling points condense at lower trays, whereas the fractions having lower boiling points condense at higher trays (Figure IV.2). The gasoline obtained by such fractional distillation is called **straight-run gasoline.**

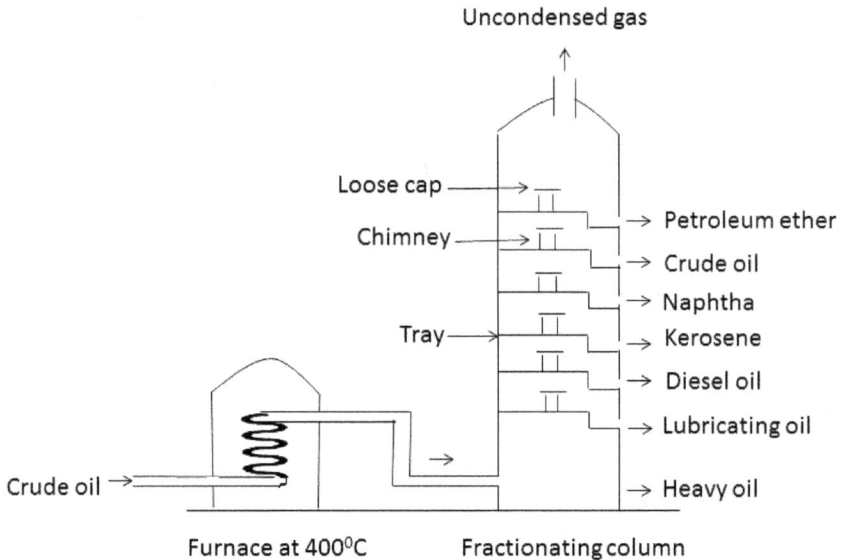

Figure IV.2 Fractional distillation process.

IV.3.3 Synthetic Petrol

The petroleum obtained from the fractional distillation of basic petroleum oil is called as straight line petrol. As the use of gasoline is increasing significantly these days, the amount of straight run gasoline is not sufficient to meet the requirements. Hence, there is a need to develop optimal process of synthesizing petrol, which can help to fulfil the current fuel needs.

Manufacture of synthetic petrol (or) hydrogenation of coal

❖ Coal contains about 4.5% hydrogen compared to 18% in petroleum. So, coal is a hydrogen deficient compound.

❖ Gasoline is produced when coal is heated with hydrogen at high temperature under high pressure.

❖ The production of liquid fuels from solid coal through hydrogenation is called hydrogenation of coal or synthetic petrol.

An improved method available for the hydrogenation of coal is Bergius process.

Bergius Process (or) direct method

The principle and procedure of the Bergius process to synthesize synthetic petrol are described below.

Principle
In this process, hydrogen reacts with coal at high temperature to form saturated hydrocarbons, which decompose to produce mixture of lower hydrocarbons.

Procedure
In this method, the finely powdered coal is prepared into a paste with heavy oil and a catalyst residue is mixed with it. The paste is pumped down with hydrogen gas into the converter, where the paste is heated to 400-450 °C under a pressure of 200-250 atm (Figure IV.3).

During this process, hydrogen combines with coal to form saturated higher hydrocarbons, which undergo subsequent decomposition at higher temperature to generate a mixture of lower hydrocarbons. This lower hydrocarbon mixture is condensed to produce crude oil. As the last step, the crude oil is subjected to fractionation process to yield fractions as i) gasoline, ii) middle oil and iii) heavy oil.

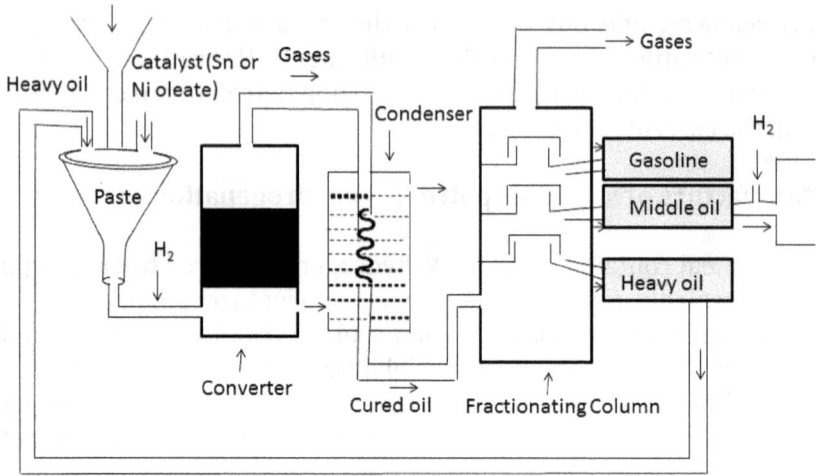

Figure IV.3 Bergius process.

IV.3.4 Knocking

Knocking is a type of explosion due to quick pressure increase taking place in an internal combustion (IC) engine. In a petrol engine, a combination of gasoline vapour and air at 1:7 ratio is used as fuel. This combination is compressed and ignited by an exciting spark. The yield of oxidation reaction increases the pressure and pushes the piston down the cylinder. During this process, the rate of explosion of the fuel will increase and the depleting portion of the fuel-air combination gets ignited, which produces an unstable sound known as knocking. Knocking property of the fuel reduces the efficiency of the engine. Thus, the gasoline fuel must have the characteristic to resist knocking.

Improvement of anti-knock characteristics

❖ Combination petrol of high octane number (see below) with petrol of low octane number, so that the octane number of the mixture can be improved.
❖ The addition of anti-knock agents such as tetraethyl lead (TEL).
❖ Addition of aromatic phosphates to petrol as anti-knock agent to avoid lead pollution due to TEL.

Octane number

The most common way of expressing the knocking characteristics of a combustion engine fuel is by octane number introduced by Edger in 1972. It has been found that n-heptane (CH_3-CH_2-CH_2-CH_2-CH_2-CH_2-CH_3) knocks badly and hence, its anti-knock value has been arbitrarily given zero. It is also observed that iso-octane exhibits very little knocking, hence, it has been given a value of 100.

	Anti-knocking value
CH_3-CH_2-CH_2-CH_2-CH_2-CH_2-CH_3	0
2,2,4-trimethylpentane (isooctane)	100

Thus, octane number (or rating) of a gasoline (or any other engine fuel) is the percentage of iso-octane present in an iso-octane and n-heptane mixture. Thus, if a sample of petrol exhibits as much of knocking as a mixture of 75 parts of iso-octane and 25 parts of n-heptane, then its octane number is taken as 75.

Fuels with octane rating greater than 100 are quite common now a days and these are rated by comparison with a blend of iso-octane with tetra ethyl lead (TEL) which greatly diminishes the knocking tendency of any hydrocarbon with which it is mixed. The value of octane number in such cases is determined by extrapolation.

Cetane number

In a diesel engine, the fuel is exploded by the application of heat and pressure instead of spark. Diesel engine fuels consist of longer chain hydrocarbons than internal combustion engine fuels. In other words, hydrocarbon molecules in a diesel fuel should be as far as possible the straight chain ones, with a minimum admixture of aromatics and side chain hydrocarbon molecules.

The suitability of a diesel fuel is determined by its cetane value which is the percentage of hexadecane present in a mixture of hexadecane and 2-methyl naphthalene.

	Cetane Number
CH_3-$(-CH_2-)_{14}$-CH_3 (hexadecane)	100
2-nethyl naphthalene	0

The cetane number of a diesel fuel can be raised by the addition of a small quantity of certain pre-ignition dopes like ethyl nitrite, isoamyl nitrite, acetone peroxide, etc. An oil of high octane number has a low cetane number and vice-versa. Consequently, petroleum crude generates petrol of high octane number and diesel of low cetane number.

IV.4 Natural Gas

Natural gas is one of the cleanest fossil fuels and is composed of lower hydrocarbons, consisting primarily of methane. Natural gas is colorless and odorless in its natural state. . It is obtained from wells dug in the oil bearing regions and coal fields. It is mainly composed of methane, ethane, propane, butane and trace amounts of other hydrocarbons along with hydrogen, carbon dioxide and monoxide.

Origin

Natural gas has been formed from microscopic plants and animals found in shallow marine environments. As this organic feedstock became buried deeper in the earth millions of years ago, heat, combined with the pressure of compaction, converted some of the biomaterial into natural gas.

Advantages of natural gas

Natural gas is a beneficial source of natural energy, offering advantages over other energy sources such as fewer impurities, less chemical complexity, lower degree of pollution as well as lower production of carbon dioxide (which is the primary greenhouse gas), sulfur dioxide (primary precursor of acid rain), nitrogen oxides (primary precursor of smog) and particulate matter (affects health and visibility) than other energy sources. Natural gas is used for manufacturing fertilizer, generating power for homes and businesses, transportation on land and sea, electricity generation, etc.

IV.4.1 Compressed Natural Gas

Natural gas obtained along with petroleum in oil wells is called 'wet gas'. As the natural gas is compressed, it is termed as compressed natural gas (CNG). The compressed natural gas is stored in tanks at a pressure of 3,000 or 3,600 psi. The primary component present in CNG is methane. Natural gas engines have a higher efficiency compared to petrol engines due to higher octane number of natural gas than petrol. One should note that CNG is different from liquefied natural gas (LNG). Though both are natural gases, the major difference is their physical state. CNG is compressed form of natural gas (gas phase), while LNG is the liquefied form of gas (liquid phase). CNG has a lower cost of production and storage compared to LNG because LNG requires expensive cooling processes and cryogenic tanks etc.

Characteristics of CNG

* CNG is the clean alternative fuel to petroleum products.
* CNG is one of the cheapest fuels, less expensive than petrol and diesel.
* CNG produces less carbon monoxide and hydrocarbon (HC) emission, thus, is an environmental friendly fuel.
* CNG requires more air for ignition and its ignition temperature is about 550 °C.

IV.4.2 Liquefied Petroleum Gas (LPG)

LPG is obtained during fractional distillation of crude petroleum oil as a byproduct. LPG comprises of lower hydrocarbons such as propane and butane. These gaseous molecules can be converted to liquid under pressure, therefore, the liquefied fuel can be easily stored in cylinders and transported. LPG contains

* n-Butane-38.5%
* Iso butane-37%
* Propane-24.5%

Advantages of LPG over gaseous fuels

* LPG consists of hydrocarbons such as propane and butane, thus, it burns completely with no residue formation.
* LPG is characterized by high thermal efficiency and heating rate.
* LPG is simple to manipulate.
* It is less hazardous than gaseous fuels.

❖ The LPG has higher calorific value than the other gaseous fuels containing H_2 or CO. The calorific value of LPG is 7 fold higher than coal gas and 3 fold than natural gas.

Disadvantages of LPG over other gaseous fuels

❖ It has faint odour, leakage cannot be simply detected, thus, requiring additives for detection.
❖ It has lower octane value.
❖ It should be handled in high pressure.

LPG is highly inflammable. It can exist as a gas under atmospheric pressure, but can be readily liquefied under pressure for easy transportation and storage. It is cheap, has high knocking-resistance and burns clearly without leaving any residue, thus, leading to its use as most common domestic and industrial fuel. It has a calorific value of 28000 kcal/kg. It is colorless and odourless. To enhance safety, a strong smelling substance (ethyl mercaptan (C_2H_5SH)) is commonly added to LPG cylinders for the early detection of gas leakage.

Table IV.1 also presents a comparison of the physical properties of CNG and LPG.

Table IV.1 Difference between CNG and LPG

Properties	CNG - Methane	LPG - Propane
Chemical formula	CH_4	C_3H_8
Energy content	9 MJ/L	25 MJ/L
Storage pressure	20-25 MPa	2 MPa
Air:gas combustion ratio	10:1	25:1
Operating pressure	1.1 kPa	2.75 kPa
Density (vs air)	.5537:1	1.5219:1
Cylinder weight	1	≈3x
State	Gas	Liquid or Gas

IV.5 Power Alcohol and Biodiesel

IV.5.1 Power Alcohol

In order to provide anti-knocking property, ethyl alcohol is generally mixed with petrol in the range 5-25% to use as fuel in internal combustion engine. Owing to this, ethyl alcohol is called as "power alcohol".

Advantages of power alcohol

- ❖ Anti-knocking property of ethyl alcohol is higher than petrol as evidenced from its higher octane number (ethyl alcohol 90, petrol 65).
- ❖ The ethyl alcohol is added to petrol to increase the octane number of petrol.
- ❖ Ethyl alcohol absorbs water present in the petrol.
- ❖ Ethyl alcohol contains oxygen atom, which also aids in complete combustion of fuel. In addition, it also minimizes the emission of CO, particulate matter and hydrocarbon.
- ❖ Power alcohol is cheaper than petrol.

Disadvantages of power alcohol

- ❖ Ethyl alcohol has lower calorific value than petrol (7000 cal/g for ethyl alcohol and 11500 cal/g for petrol).
- ❖ Ethyl alcohol reduces power output of fuel upto 35%.
- ❖ Atomization of ethyl alcohol is difficult at lower temperature which causes starting trouble.
- ❖ Ethyl alcohol may convert into acetic acid by oxidation leading to corrosion of engine parts.
- ❖ Since ethyl alcohol contains oxygen atoms, which help for complete combustion of fuel. Hence, the power alcohol consumes less amount of air for combustion which requires modification of carburetor and engine design.

IV.5.2 Biodiesel

Fuels derived from plant seeds and vegetables are known as biodiesel. Vegetable oil includes 90-95% triglycerides through the small quantities of diglycerides, fatty acids, phospholipids, etc. (triglycerides are esters of long chain fatty acids, similar to stearic acid and palmitic acid). Biodiesel is used in aircrafts, automobiles and as heating oil.

Advantages of biodiesel

- ❖ Biodegradable
- ❖ It is readily obtained from renewable resources, which are abundantly available
- ❖ Lower extent of gaseous pollutants as compared to the conventional diesel fuel
- ❖ It can be produced from vegetable oils of different molecular weights and other properties
- ❖ Excellent engine performance and less burn production are achieved

Disadvantages of biodiesel

- ❖ The biomaterials are hygroscopic, thus, biodiesel can absorb water from atmosphere
- ❖ It may decrease the horse power of the engine
- ❖ Biodiesel has about 10% higher nitrogen oxide production than conventional petroleum
- ❖ Since the viscosity of vegetable oil is high, atomization is difficult, thus, leading to deficient burning
- ❖ Oxidation and thermal polymerization of vegetable oils cause deposits in the engine
- ❖ The use of vegetable oils as straight fuel requires adjustment of the conventional diesel engine design

Production of Biodiesel

The direct use of vegetable oils in engines leads to several problems such as clogging of fuel filter, poor atomization and incomplete combustion. Transformation of vegetable oils into biodiesel can be achieved using (i) heating/pyrolysis, (ii) dilution/blending, (iii) micro-emulsion and (iv) transesterification. Among the above techniques, transesterification is an extensive, convenient and the most promising method for the reduction of viscosity, density and other properties of raw vegetable oils for effective use in internal combustion engines.

Transesterification

The vegetable oils are composed of triglycerides of fatty acids, which are esters containing three fatty acid molecules and a trihydric alcohol (glycerol). In this process, an alcohol reacts with the fatty acids, in presence of a strong base, to form the mono-alkyl ester or biodiesel and crude glycerol. Commonly, ethanol or methanol are used and is base catalyzed by either potassium or sodium hydroxide.

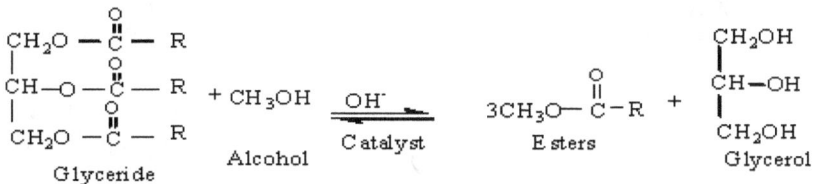

After the transesterification reaction, separation of the crude glycerol phase and light biodiesel phase is carried out. The crude biodiesel also requires some purification prior to use.

IV.6 Combustion of Fuels

Fuel refers to any substance which on effective burning produces large amount of heat and light, through the process of combustion. The heat and light produced from the fuels can be used economically for domestic and industrial purposes.

Combustion refers to the rapid oxidation of fuel accompanied by the production of heat, or heat and light. Complete combustion of a fuel is possible only in the presence of an adequate supply of oxygen.

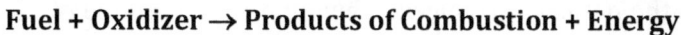

Fuel + Oxidizer → Products of Combustion + Energy

Rapid fuel oxidation results in a large amount of heat release. Solid or liquid fuels must be vaporised to gas to burn. Heat is required to vaporise liquids or solids into gases. Fuel gases will burn in their normal state, if enough air is present.

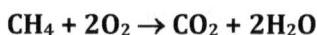

$$CH_4 + 2O_2 \rightarrow CO_2 + 2H_2O$$

IV.6.1 Calorific Value

Calorific value of a fuel is "the total quantity of heat liberated, when a unit mass (or volume) of the fuel is burnt completely".

Units of heat

- ❖ Calorie is defined as the amount of heat required to raise the temperature of one gram of water through one degree Centigrade (15-16 °C).
- ❖ Kilocalorie is defined as 'the quantity of heat required to raise the temperature of one kilogram of water through one degree Centigrade. Thus, 1 kcal = 1,000 cal.
- ❖ British thermal unit (B.T.U.) is defined as the quantity of heat required to raise the temperature of one pound of water through one degree Fahrenheit (60-61 °F). This is the English system unit. 1 B.T.U. = 252 cal = 0.252 kcal; 1 kcal = 3.968 B.T.U.

Higher or gross calorific value (HCV)

Higher or gross calorific value is defined as the total amount of heat produced when a unit mass/volume of the fuel has been burnt completely and the products of combustion have been cooled to room temperature.

It is explained that during the determination of the calorific value of hydrogen containing fuels experimentally, the hydrogen is converted into steam and as it condenses at room temperature, the latent heat of condensation of steam also gets included in the measured heat which is called gross or higher calorific value (HCV).

Lower or net calorific value (LCV)

LCV is the net heat produced, when unit mass/volume of the fuel is burnt completely and the products are permitted to escape.

It is explained that the steam (water vapour) formed during combustion is not condensed and allowed to escape along with hot combustion gases. Hence, a lesser amount of heat is measured.

Net calorific value = Gross calorific value - Latent heat of condensation of water vapour produced

= GCV - **Mass of hydrogen per unit weight of the fuel burnt x 9 x latent heat of condensation of water vapour**

Theoretical calculation of calorific value

Carbon, hydrogen and sulphur in the fuel combine with oxygen in the air to form carbon dioxide, water vapour and sulphur dioxide, releasing 8084 kcal, 28922 kcal and 2224 kcal of heat respectively.

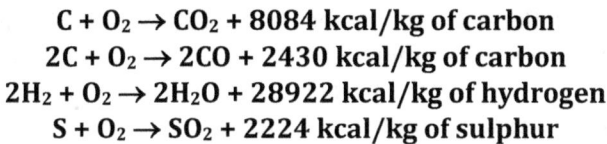

$$C + O_2 \rightarrow CO_2 + 8084 \text{ kcal/kg of carbon}$$
$$2C + O_2 \rightarrow 2CO + 2430 \text{ kcal/kg of carbon}$$
$$2H_2 + O_2 \rightarrow 2H_2O + 28922 \text{ kcal/kg of hydrogen}$$
$$S + O_2 \rightarrow SO_2 + 2224 \text{ kcal/kg of sulphur}$$

Oxygen is present in combined form with hydrogen in the form of fixed hydrogen. Thus,

Amount of hydrogen available for combustion = [(Total mass of hydrogen in fuel) - (fixed hydrogen)]
= Total mass of hydrogen in fuel – 1/8th mass of oxygen in the fuel (8 parts of oxygen combines with 1 part of hydrogen to form water)

Delong's formula (for calculation of calorific value):

HCV = 1/100 (8080 C + 34500 (H – O/8) + 2240 S) kcal/kg
where C, H, S and O are the percentages of each in the fuel.

Oxygen is assumed to be present in the combined state with hydrogen as water.

LCV = [HCV – (9/100) × H × 587] kcal/kg

Example

Calculate the gross (GCV) and net calorific (LCV) value of a coal sample having the following composition:
C = 80%, H = 7%, O = 3%, S = 3.5%, N = 2.5% and ash 4.4%

Solution
G.C.V = 1/100 (8080 x %C + 34500(%H - %O/8) +2240 x %S)
       = 1/100 (8080 x 80 + 34500 (7 - 3/8) + 2240 x 3.5)

$$= 8828.0 \text{ kcal/kg}$$

Net calorific value = G.C.V - (0.09H x 587)
$$= 8828 - (0.09 \ x \ 7 \ x \ 587)$$
$$= 8458.2 \text{ kcal/kg}$$

IV.6.2 Ignition Temperature (Spontaneous Ignition Temperature)

❖ It is defined as "the lowest temperature at which a combustible substance when heated catches fire in air and continues to burn without the presence of a spark or flame".

❖ The necessary condition to achieve ignition is to supply the required temperature to a combustible mixture.

❖ Minimum of ignition energy (MIE) is one of the parameters which characterizes the explosion properties of a fuel. MIE depends on the nature of fuel, mixture content and conditions of ignition.

❖ Ignition delay τ_{in} is a time interval (delay) after which the pressure rise of explosion is observed as an effect of burning. For conventional fuels (e.g. gasoline), the ignition delay is in the range of 20-40 ms.

IV.6.3 Explosive Range

❖ The minimum concentration of a combustible gas or vapour required to support its combustion in air is known as its lower explosive limit (LEL). Below this level, the mixture will not burn.

❖ The highest concentration of a gas or vapour which will burn in air is known as the upper explosive limit (UEL).

❖ The range between the LEL and UEL is known as the flammable range for that gas or vapour.

For a fire or explosion to occur, a combustible gas and oxygen (air) must exist in certain proportions, along with an ignition source, such as a spark or flame. The ratio of fuel and oxygen that is required varies with each combustible gas or vapour.

Lower and upper explosive limit values of some common organic solvents are also demonstrated in Table IV.2 below. As can be observed, a large range of LEL and UEL values exist for these solvents.

Table IV.2 LEL and UEL of some organic solvents

	LEL	UEL
Acetone	2.6	13.0
Benzene	1.3	7.4
Butane	1.8	8.4
Ethane	3.0	12.4
Ethanol	3.3	19.0
Gasoline	1.2	7.1
Hexane	1.2	7.4
Hydrogen	4.0	75.0
Methane	5.0	15.0
Methanol	6.7	36.0
Pentane	1.4	7.8
Propane	2.1	9.5
Toluene	1.2	7.1
Xylene	1.1	6.6

IV.7 Flue Gas Analysis

The gaseous byproducts released during combustion of fuel are known as flue gas. Commonly, the mixture of SO_2, CO_2, O_2 and CO gases is released from a combustion chamber when a fuel is burnt.

The combustion products are mainly gaseous in nature. The analysis of gaseous byproducts provides an idea of combustion process of the particular fuel. The analysis also conforms whether a combustion process is complete or not. For example, the presence of CO in the flue gas indicates incomplete combustion which in turn indicates that the supply of oxygen is not sufficient in the process.

Analysis using ORSAT apparatus

The flue gas analysis is carried out by using Orsat apparatus. The analysis of flue gas generally deals with the determination of CO_2, O_2 and CO by absorbing them in the respective solution of KOH, alkaline pyrogallol and ammonium cuprous chloride.

An Orsat apparatus consists of the following: (i) a burette, (ii) a gas cleaner and (iii) four absorption pipettes. The pipettes are interconnected by means of a manifold fitted with cocks and contain different chemicals to absorb carbon dioxide (CO_2), carbon monoxide (CO) and oxygen (O_2). The horizontal tube is connected to three different absorption bulbs for the absorption of CO_2, O_2 and CO respectively. The lower end of the burette is connected to the levelling bottle by means of rubber tube. The level of water in the levelling bottle can be raised or lowered by raising or lowering the water reservoir. By changing the level of water, the flue gas can be moved into various parts of the apparatus during analysis. The principle of Orsat apparatus is depicted in Figure IV.4.

It is essential to follow the order of absorbing the gases: CO_2 first, O_2 second and CO last, as the absorbent used for O_2 (i.e., alkaline pyrogallol) can also absorb some amount of CO_2 and the percentage of CO_2 left would be less.

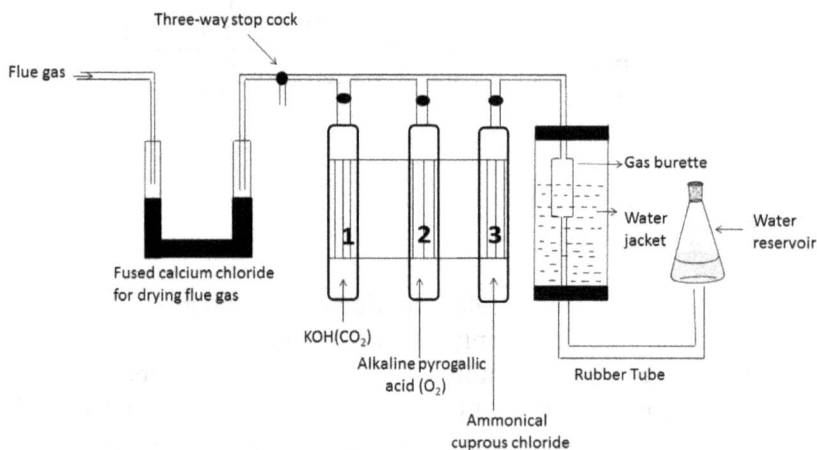

Figure IV.4 Orsat apparatus.

Absorption of CO_2

Flue gas is passed into the bulb **1** via its stopcock by raising the water reservoir. CO_2 present in the flue gas is absorbed by 50% KOH solution in distilled water. The gas is again sent to the burette and subsequently sent to bulb **1**. This process is repeated several times, by raising or lowering of water reservoir so as to ensure complete absorption of CO_2 in KOH. Now, the stopcock of bulb **1** is closed. The volume of residual gases in the burette is taken by equalizing the water level both in the burette and in the water reservoir. The difference between original volume and the volume of the gases after passing through KOH provides the volume of CO_2 absorbed during the process.

Absorption of O_2

Stopcock of bulb **1** is closed and bulb **2** is opened. Oxygen present in the flue gas is absorbed by alkaline pyrogallol (200 g KOH and 25 g pyrogallol are dissolved in 500 mL distilled water). The absorption process is same as in bulb **1**.

Absorption of CO

Finally, the stopcock of bulb **2** is closed and stopcock of bulb **3** is opened. Carbon monoxide present in the flue gas is absorbed by ammoniacal cuprous chloride (100 g Cu_2Cl_2 dissolved in a mixture of 125 mL liquid NH_3 and 375 mL distilled water) in bulb **3**. The absorption process is also same in this case as in bulb **1**. Since the total volume of the gas taken for analysis is 100 mL, the volume of the constituent is its percentage. The residual gas after the above three determinations is taken as nitrogen. Further, as the content of CO in the flue gas is generally very low, it should be measured carefully.

UNIT V

ENERGY SOURCES AND STORAGE DEVICES

V.1 Introduction

We essentially depend on the conventional sources for energy production, such as coal, petroleum, natural gas, nuclear energy, etc. These sources are limited in quantity and can be exhausted in near future due to their continuous and rapid use. To meet the energy needs of the future, the scientists have accelerated the research on use of non-conventional sources of energy such as wind energy, solar energy, tidal energy, etc.

V.2 Nuclear Fission

The U^{235} is bombarded by thermal neutron (low energy neutron), which results in its splitting into two approximately identical parts with the liberation of a large amount of energy.

V.2.1 Definition

Nuclear fission is defined as "the process of splitting of heavier nucleus into two (or) more smaller nuclei with simultaneous liberation of a large amount of energy".

V.2.2 General Mechanism of Nuclear Fission

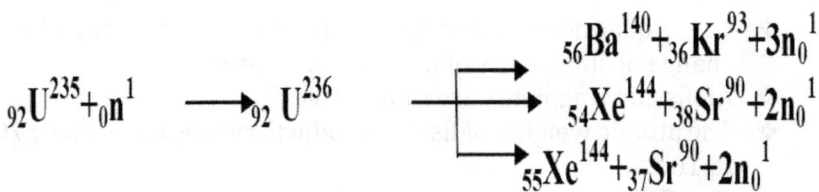

$$_{92}U^{235} + _0n^1 \longrightarrow _{92}U^{236} \begin{cases} \longrightarrow _{56}Ba^{140} + _{36}Kr^{93} + 3n_0^1 \\ \longrightarrow _{54}Xe^{144} + _{38}Sr^{90} + 2n_0^1 \\ \longrightarrow _{55}Xe^{144} + _{37}Sr^{90} + 2n_0^1 \end{cases}$$

As U^{235} is bombarded by thermal neutron (slow moving), it produces unstable U^{236}. The unstable U^{236} splits into two approximately identical nuclei with the liberation of neutrons and large amount of energy.

V.2.3 Illustration

Splitting of U[235] has been shown below in Figure V.1.

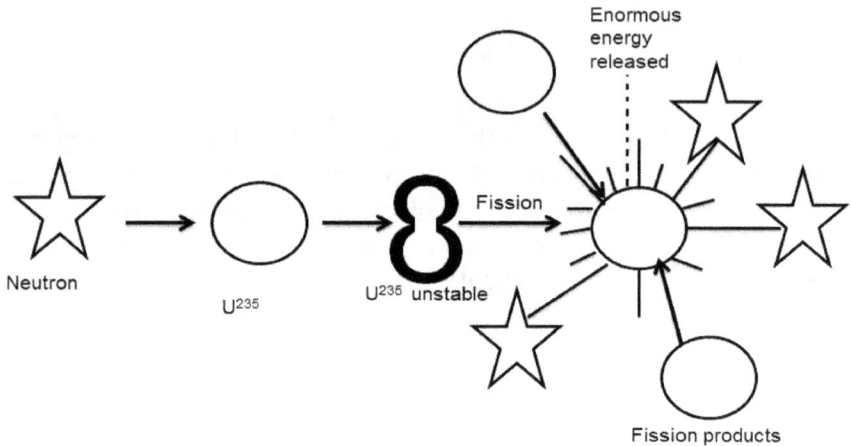

$$\,^{1}_{0}n + \,^{235}_{92}U \rightarrow \,^{141}_{56}Ba + \,^{92}_{36}Kr + 3\,^{1}_{0}n$$

Figure V.1 Nuclear fission reaction.

V.2.4 General Characteristics of Nuclear Fission Reaction

❖ A heavy nucleus (U[235] or Pu[239]), when bombarded by slow energy moving neutrons, split into two or more nuclei.

❖ Two or more neutrons are twisted by fission of every nucleus.

❖ Large quantities of energy are produced as a result of exchange of small mass of nucleus into energy.

❖ All fission fragments are radioactive.

❖ The atomic weights of fission products ranges from about 70 to 160.

❖ The fission reactions are identical propagating chain reactions because fission products surround neutrons, which act as further source of fission in additional nuclei.

❖ The nuclear chain reactions can be controlled and maintained gradually by absorbing a preferred number of neutrons. This method is used in the nuclear reactor.

❖ The every secondary neutron released in the fission process does not bombard another nucleus, thus, a small quantity escapes into air and, hence, a chain reaction cannot be controlled.

V.2.6 Advantages of Nuclear Fission Energy over Fossil Fuel Energy

❖ A small amount of nuclear fuel (U^{235}) generates a large amount of energy as heat. On the other hand, large quantity of fossil fuel is essential to generate significant amount of heat.

❖ Nuclear energy is a clean energy source in terms of carbon emissions.

❖ Nuclear energy is a reliable source of power because the nuclear reactors used today have a long life (normal service life of reactors run in decades).

❖ Nuclear energy is a very powerful and efficient source of power (nuclear power has the highest energy density known).

❖ A nuclear power plant generates large amount of clean electricity in a relatively small space.

V.2.7 Disadvantages of Nuclear Fission Energy over Fossil Fuel Energy

❖ The nuclear fission processes can lead to serious pollution problems.

❖ The largest challenge of nuclear fission energy is the safe disposal of nuclear waste. No such problem is faced in the disposal of fossil fuel.

❖ Nuclear power produces nuclear waste which is hazardous for the environment.

❖ Nuclear power is not a renewable energy source because uranium, plutonium and thorium are finite resources.

❖ Disposing nuclear waste is an expensive activity which requires both high expenses and heavy military guard for the objective.

❖ Nuclear reactors have already resulted in several radioactive disasters which have affected both human life and environment.

V.3 Nuclear Fusion Reaction

$$_1H^2 + {}_1H^2 \longrightarrow {}_2He^4 + energy$$

$$_1H^2 + {}_1H^2 \longrightarrow {}_2He^4 + {}_0n^1$$

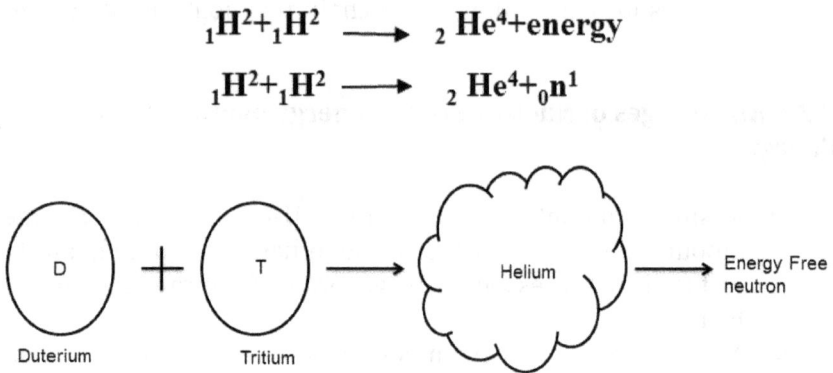

Figure V.2 Nuclear fusion reaction.

The nuclear fusion (Figure V.2) is defined as "the process of combination of lighter nuclei into heavier nuclei, with simultaneous liberation of large amount of energy".

V.3.1 Characteristics of Nuclear fusion

❖ The nuclear fusion is possible only if the distance between the nuclei is in the range of one Fermi.
❖ The energy released in a fusion reaction is four times more compared to the energy released in fission reaction.
❖ The sufficient amount of kinetic energy has to be provided to make a fusion reaction possible.
❖ As electrostatic repulsion increases with increase in atomic number of nuclei, only lighter nuclei can undergo nuclear fusion reaction.

V.3.2 Illustration

The fusion reaction is continuously happening in stars and sun, where hydrogen is converted into helium. The fusion reactions would occur only when the atoms are heated to extraordinarily high temperatures. Also, the nuclei are positively charged, thus, they must attain high speed to overcome the electrostatic repulsion. In fact, the hydrogen to helium fusion reaction in the sun occurs at about 100 million °C. Thus, to initiate a fusion reaction, temperature of high order must be reached. For this reason, the fusion reactions are also called thermonuclear reactions. In comparison, a high temperature is generated by a fission reaction.

V.3.3 Differences Between Nuclear Fission and Fusion

Nuclear fission	Nuclear fusion
It is the process of breaking of heavier nuclei into lighter nuclei.	It is the process of combination of lighter nuclei.
It emits radioactive rays.	It does not emit any radioactive rays.
It occurs at normal temperature.	It occurs at high temperature (> 10^6 K).
The mass number and atomic number of new elements are lower than that of parent nucleus.	The mass number and atomic number of product are higher than that of starting elements.
It gives rise to a chain reaction.	It does not give rise to chain reaction.
It emits neutrons.	It emits positrons.
It can be controlled.	It cannot be controlled.

V.4 Nuclear Chain Reactions

The nuclear fission reaction of one U^{235} atom produces three neutrons with other products. The three neutrons subsequently bombard another three U^{235} nuclei. Therefore, each fission reaction produces three neutrons which initiate fission of U^{235} nuclei. Thus, a chain reaction of nuclear fission reactions takes place, as shown in Figure V.3, with a release of large amount of energy. Therefore, the fission reaction of U^{235} by slow moving neutrons is a nuclear chain reaction.

V.4.1 Definition

The fission reaction where the neutrons from the earlier step continue to spread and carry out fission reaction again is called nuclear chain reaction.

V.4.2 Criterion for Nuclear Chain Reaction

In order to maintain a nuclear chain reaction, adequate amount of U^{235} must be present to utilize the neutrons, otherwise neutrons will break away from the surface.

V.4.3 Critical mass

The smallest amount of fissionable material (U^{235}) essential to carry on the nuclear chain reaction is called critical mass. The critical mass of U^{235} lies between 1 kg to 100 kg.

Super-critical mass: If the nuclear mass of the fissionable material (U^{235}) is more than the critical mass, it is called super-critical mass.

Sub-critical mass: If the nuclear mass of the fissionable material is smaller than the critical mass, it is called sub-critical mass.

The nuclear masses greater or lesser than the critical mass shall hinder the propagation of the chain reaction.

V.4.4 Illustration

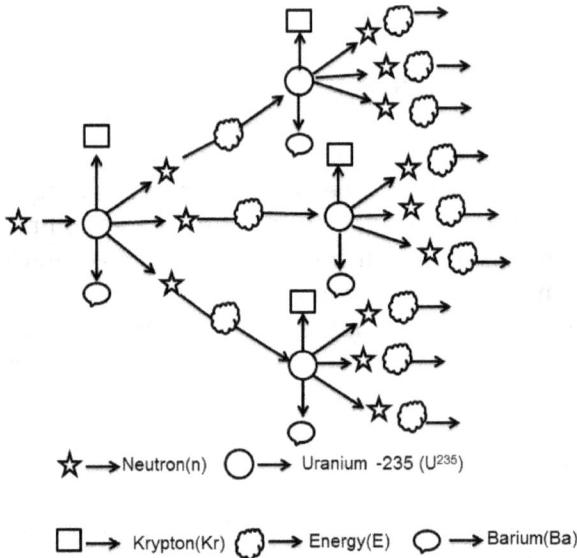

Figure V.3 Nuclear chain reaction.

As mentioned before (and in Figure V.3), as U^{235} nucleus is hit by a thermal neutron, it undergoes the fission reaction with the liberation of three neutrons. The three neutrons, liberated in the process, strike additional three U^{235} nuclei causing nine successive reactions. These nine reactions further give rise to 27 reactions. This propagation of the reaction by multiplication for each fission reaction is called nuclear chain reaction.

V.5 Nuclear Energy

The large amount of energy released for the duration of the nuclear chain reaction of heavy isotope like U^{235} or Pu^{239} is called nuclear energy.

V.5.1 Definition

The energy released by the nuclear fission or nuclear fusion is called nuclear energy (Figure V.4).

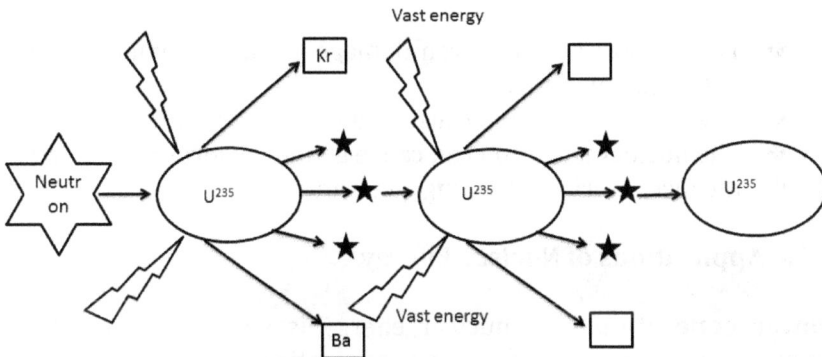

Figure V.4 Generation of nuclear energy.

The fission of U^{235} or Pu^{239} occurs immediately, producing enormous amount of energy in the form of heat and radiation.

V.5.2 Source of the Release of Energy

The large amount of energy released for the duration of the nuclear fission is due to the loss in mass, as soon as the reaction takes place.

It has been experimentally observed that during nuclear fission, the sum of the masses of the products formed is slightly less than the sum of the masses of the target species and bombarding neutron. The loss in mass gets converted into energy according to Einstein equation $E = \Delta mc^2$

$$\therefore \Delta m = (M - M')$$

where,
c = velocity
Δm = loss in mass
E = energy
M = mass of radioactive substance before emitting radiation
M' = mass of radioactive substance after emitting radiation

V.5.3 Hazards of Nuclear Energy

The radiation is damaging to the living organisms. The long and invariable exposure to these radiations can have the following hazards:

- ❖ The nuclear radiation can damage the arrangement of cells in the human body.
- ❖ The nuclear radiation causes cancer and blindness.
- ❖ The nuclear radiation can cause genetic disorder in humans.
- ❖ It causes sterility in young generation.

V.5.4 Applications of Nuclear Energy

Power generation: The nuclear energy is an environmental responsive energy resource for power generation.

Resource of pure water: The discharged water from the nuclear reactor is free from radiation and is adequate to conserve animals and aquatic animals.

Physical care: Radioactive isotopes find use in cancer cure by radiotherapy. It is also used for sterilization to destroy microorganisms.

Agriculture: It is used to control agricultural pests. Nuclear radiation delays ripening of fruits.

V.6 Types of Nuclear Fission Reactions

Nuclear fission reactions are of two types:

- ❖ Uncontrolled nuclear fission reaction
- ❖ Controlled nuclear fission reaction

V.6.1 Uncontrolled nuclear fission reaction

If a nuclear fission reaction is prepared to occur in an uncontrolled manner, the released energy can only be used for critical purposes.

V.6.2 Controlled nuclear fission reaction

If a nuclear fission reaction is prepared to occur in a controlled manner, the released energy can be used for several constructive purposes.

The nuclear reactors are able to control the nuclear chain reactions by absorbing the neutrons excluding one neutron using appropriate resources like B or Cd. The excluded neutron is available for propagating the fission reaction further. Such a reaction is called controlled nuclear fission reaction and the energy release is effectively controlled.

V.7 Nuclear Reactors

The reactors planned to carry out nuclear reactions for generation of heat, electricity and radioactive isotopes are called nuclear reactors.

V.7.1 Definition

The arrangement or apparatus used to carry out fission reaction under controlled conditions is called a nuclear reactor.

V.7.2 Light Water Nuclear Power Plant

The nuclear power plant in which U^{235} fuel rods are submerged in water, is termed as light water nuclear plant. Here, water acts as both coolant and moderator.

Components of a light water nuclear power plant

The main components of the light water nuclear power plant are:

Fuel rods: The fissionable materials used in the nuclear reactor is enriched U^{235}. The enriched fuel is in the form of rods or strips.
Purpose: It produces heat energy along with neutrons which initiate the nuclear chain reaction.

Control rods: To control the fission reaction (rate), movable rods, made of cadmium, are balanced between fuel rods. These rods can be lowered or raised and control the fission reaction by absorbing excess neutrons.

   If the rods are deeply inserted inside the reactor, these will absorb more neutrons and the reaction becomes very slow. On the other hand, if the rods are pressed outwards, these will absorb less neutrons and the reaction will be very fast.
 Function: It controls the nuclear chain reaction and avoids the damage to the reactors.

Moderators: The substances used to slow down the neutrons are called moderators. As the fast moving neutrons collide with moderator, these lose energy and become slow.
Function: The kinetic energy of fast moving neutrons (1 meV) is reduced to 0.25 eV.

Coolants: In order to absorb the heat during fission, a liquid called coolant is scattered in the reactor core. It enters the base of the reactor and leaves at the top. The heat carried by outgoing liquid is used to generate steam.

Pressure vessel: It encloses the core and provides the entrance and exit passages for coolant.

Protective shield: The nuclear reactor is enclosed in a thick concrete shield (more than 10 meters thick).
Function: The environment and operating personnel are confined from destruction in case of leakage of radiation.

Turbine: The steam generated in the exchanger is used to operate a steam turbine, which drives a generator to produce electricity.

Functioning

The fission reaction is controlled by inserting or removing the control rods of B10 mechanically from the places in between the fuel rods. The heat emitted by the fission of U^{235} in the fuel core is absorbed by the coolant (light water). The heated coolant (water at 300 °C) subsequently goes to the heat exchanger containing sea water. The coolant transfers the heat to sea water, which is transformed into steam. The generated steam drives the turbines, thereby, generating electricity (Figure V.5).

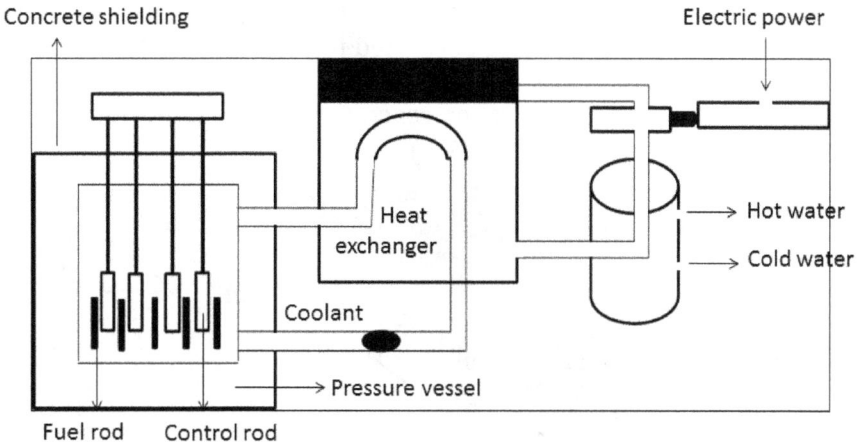

Figure V.5 Functioning of light water nuclear power plant.

Pollution

Though nuclear power plants are important for the production of electricity, these pose a serious hazard to environment. Removal of reactor waste is a serious problem since the fission products viz., Ba^{139} & Kr^{92} are radioactive. These emit hazardous radiation for several hundred years. Thus, the waste is packed in material barrels, which are embedded deep in the sea. Salt water does not allow the radiation to be emitted out.

V.7.3 Breeder Reactor

The breeder reactor is the reactor which converts non-fissionable material (U^{238}, Th^{232}) into fissionable material (U^{235}, Pu^{239}), consequently the reactor manufactures or breeds more fissionable material than it consumes.

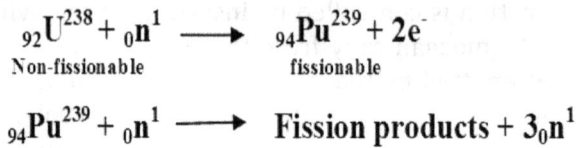

$$_{92}U^{238} + {}_0n^1 \longrightarrow {}_{94}Pu^{239} + 2e^-$$
Non-fissionable   fissionable

$$_{94}Pu^{239} + {}_0n^1 \longrightarrow \text{Fission products} + 3{}_0n^1$$

Illustration

In the breeder reactor, only one neutron out of the three emitted in the fission of U^{235} is used to propagate the fission chain reaction. The additional two neutrons are permitted to counter with U^{238}. Consequently, two fissionable atoms of U^{235} are generated. Therefore, as mentioned above, the breeder reactor produces more fissionable substance than it uses (Figure V.6).

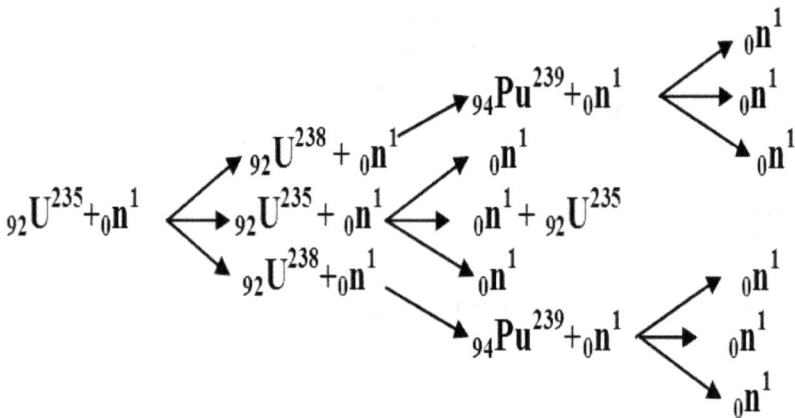

Figure V.6 Illustration of a reaction in breeder reactor.

V.8 Solar Energy Conversion

Solar energy conversion is the process of transformation of direct sunlight into more useful forms. The solar energy conversion occurs by the following two mechanisms:

- ❖ Thermal conversion
- ❖ Photoelectric conversion

V.8.1 Thermal Conversion

Solar energy is an essential source for low-temperature heat which is helpful for heating buildings, water and refrigeration.

Types of thermal conversion

- ❖ Solar heat collectors
- ❖ Solar water heaters

Solar heat collectors

The solar heat collectors consist of ordinary materials similar to bricks and stones which can absorb heat for the duration of the sunlight time and discharge it gradually at night.

Solar heat collectors are commonly used in cold places, where houses are maintained in warm conditions using solar heat collectors.

Solar water heaters

A solar water heater is a device that can be employed for using sunlight to heat the water for domestic purposes such as baths, showers, etc. It consists of

- ❖ a **thermal panel** (solar collector) installed on the roof
- ❖ a **tank** to store hot water
- ❖ **accessories**, such as a **circulating pump** to carry the solar energy and a **thermal regulator**.

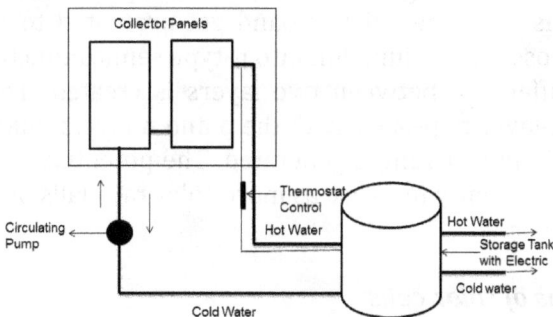

Figure V.7 Solar water heater.

V.8.2 Photo-conversion

Photo-conversion involves conversion of light energy directly into electrical energy by the photoelectric effect.

The solar radiation falls on the surface of a metal, it is absorbed and used to excite and eject electrons from the surface. The ejected electrons move in a circuit and produce electric current.

Photogalvanic cell (solar cell)

Definition

Solar energy is directly converted into electrical energy.

Principle

The principle in the solar cells is based on the photovoltaic effect. The solar rays plunge on two layered semi-conductor devices and a potential difference between the two layers is produced. This potential difference causes flow of electrons and produces electricity.

Construction

The solar cells consists of a p-type semiconductor (for example, Si doped with B) and n-type semiconductor (for example Si doped with P) in close contact with each other.

Operation

The sun radiation plunges on the top layer of p-type semiconductor, the electrons from the valence band get promoted to the transfer band and cross the p-n junction into n-type semiconductor. Thereby, potential difference between two layers is created. The electrons flow from n-layer to p-layer with the p and n layers linked to an external circuit and current is generated. The potential difference and current consequently increase as more solar rays falls on the surface of the top layer.

Applications of solar cells

❖ Solar cells can be used to drive vehicles.

- ❖ Solar cells are useful in calculators, electronic watches and radios.
- ❖ Solar cells are used in remote areas.
- ❖ Solar cells are made of silicon are used as a source of electricity in spacecraft and satellites.
- ❖ Electrical street lights can be replaced by solar street lights.

V.9 Wind Energy

The moving air is called wind. Energy generated from the strength of wind is called wind energy. Energy possessed by the wind is due to its high speed. Thus, kinetic energy of the wind is changed into mechanical energy.

V.9.1 Types of Wind Energy Harnessing

Wind mill

The device used to convert wind energy into mechanical energy (Figure V.8).

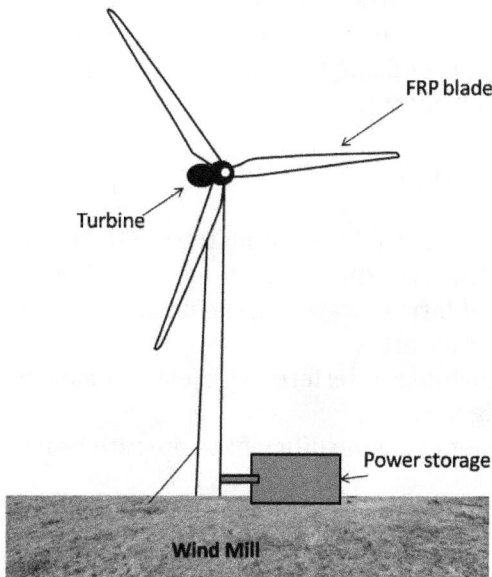

Figure V.8 A wind mill.

Structure and working of a wind mill

The wind mill consists of a wheel containing a number of blades. The wheel rotates around an axle mounted on a pole. The blowing wind rotates the wheel. One end of the axle is associated to the framework of a generator (turbine), which rotates between two poles of a physically powerful magnet. Another end of the axle is associated to the beam of the wind mill. As wings (blades) are rotated by blowing wind, these in turn rotate the turbine and electric current is produced. Thus, the kinetic energy of the wind is transformed into electric energy.

Wind fish farm

The electricity produced by a single wind mill is not sufficient for commercial purposes. In order to produce electricity on a large scale, a large number of wind mills are deployed and are connected to a single power storage. The region where large numbers of wind mills are erected to produce electricity is called wind energy farm.

V.9.2 Advantages Wind Energy

- ❖ It does not cause any pollution.
- ❖ It is very economic, after installation of wind mills, as operational cost is nil and the mills have a long service life.
- ❖ It is renewable.

V.9.3 Limitations of Wind Energy

- ❖ Public resistance for locating the wind energy farms in more populated areas due to noise and loss of visual appearance.
- ❖ The wind farms located on the traveling routes of birds are a source of hazard.
- ❖ Wind turbines interfere with electromagnetic signals (TV, radio signals, etc.).
- ❖ Wind energy is not sufficient to operate heavy machinery.

V.9.4 Uses of Wind Energy

- ❖ Wind energy is used to move the sail boats in lakes, rivers and seas.

❖ It is used to operate water pumps.
❖ It is used to run the flour mills to grind grains.
❖ It is also used to produce electricity for other purposes.

V.10 Storage Devices

Electrical energy can be stored in electrochemical cells and can be transported and reused whenever needed. The common electricity storage devices are

❖ Batteries
❖ Supercapacitors

V.10.1 Batteries

Batteries are the devices which transform chemical energy into electricity through an electrochemical reaction called oxidation-reduction (redox). Oxidation and reduction reactions are associated with the standard cell potential E°, which is calculated from the thermodynamic data as follows

$$E° = \frac{-\Delta G°}{nF}$$

where $\Delta G°$ is standard Gibbs free energy, n is number of electrons exchanged and F is Faraday constant.

Batteries can be used as a source of direct electric current. The batteries have two terminals: positive cathode (+) and negative anode (-).

Operation: As the device is switched on, the chemical reaction starts and produces electrons, which electrons travel from (-) to (+), thus, electrical work is produced.

A battery is an arrangement of several electrochemical cells connected in series.

A cell contains only one anode and cathode.
A battery contains several anodes and cathodes.

An electrochemical cell comprises:

❖ a negative electrode to which anions (negatively charged ions) migrate, i.e., anode donates electrons to the external circuit as the cell discharges
❖ a positive electrode to which cations (positively charged ions) migrate, i.e., cathode
❖ Electrolyte solution containing dissociated salts, which enable ion transfer between the two electrodes, providing a mechanism for charge to flow between positive and negative electrodes
❖ a separator which electrically isolates the positive and negative electrodes

Requirements of a battery

A useful battery should fulfill the following requirements:

❖ It should be light and compact for easy transport
❖ It should have long life both in use and out of use
❖ The voltage of the battery should not vary appreciably during its use

Types of batteries

Batteries are classified into primary (non-rechargeable) and secondary (rechargeable) depending on their capacity of being electrically recharged.

Primary batteries

In primary batteries, the electrode reactions are not reversible, i.e., applying the external energy will not reconstruct the electrodes, hence, the cells are not rechargeable. After the discharge, these are discarded. Some of the important primary batteries are briefly explained below

Zinc-carbon battery (Leclanche cell): The zinc-carbon battery is made up of zinc anode and manganese dioxide cathode. It has an electrolyte of aqueous ammonium chloride and/or zinc chloride. Powdered carbon black is used in the cathode mix to improve conductivity of the cathode. The electrochemical reactions of the zinc-carbon battery are as follows:

Aqueous NH_4Cl and $ZnCl_2$
$Zn \rightarrow Zn^{2+} + 2e^-$ (anode)
$MnO_2 + H_2O + 2e^- \rightarrow MnOOH + OH^-$ (cathode)
$Zn + 2MnO_2 + 2H_2O \rightarrow 2MnOOH + Zn(OH)_2$ (total reaction)

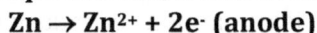

Alkaline manganese battery: The principal difference between zinc-carbon and alkaline manganese battery is the electrolyte concentrated potassium hydroxide (KOH) solution. The electrochemical reactions of the alkaline manganese battery are as follows:

Aqueous KOH solution
$Zn + 2OH^- \rightarrow ZnO + H_2O + 2e^-$ (anode)
$2MnO_2 + H_2O + 2e^- \rightarrow Mn_2O_3 + 2OH^-$ (cathode)
$Zn + 2MnO_2 \rightarrow ZnO + Mn_2O_3$ (total reaction)

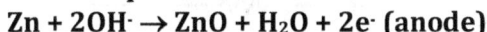

Alkaline manganese battery is more reliable and has superior performance of up to ten times of Ah capacity as compared to zinc-carbon battery at higher discharge currents.

Primary lithium battery: The electrochemical reactions of the primary lithium battery with MnO_2 cathode is given below:

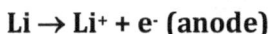

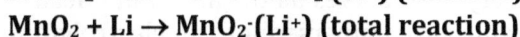

$Li \rightarrow Li^+ + e^-$ (anode)
$2MNO_2 + Li^+ + e^- \rightarrow MnO_2^-(Li^+)$ (cathode)
$MnO_2 + Li \rightarrow MnO_2^-(Li^+)$ (total reaction)

Primary lithium batteries are popular because of their high voltage as well as high energy density. In lithium batteries, lithium foils are used as anode, whereas various cathode materials such as CuO, CuS, CF, MnO_2, etc. are employed. Lithium ion conducting organic molecules or polymers are used as electrolytes. Some common primary lithium batteries are lithium/iodine battery, lithium manganese dioxide battery and lithium-lithium thionyl chloride battery.

Secondary batteries

Secondary batteries are the rechargeable batteries, i.e., the electrode reactions are reversible by applying an external voltage to reconstruct the electrodes to their original state. The storage capacity, rate of charge and discharge capability, cycling performance, environmental health and safety are few of the important characteristics of rechargeable batteries. Common secondary batteries are (i) lead-acid battery and (ii) lithium ion battery.

Lead acid battery: Lead acid battery is one of the first rechargeable batteries employed in commercial use, invented by the French physician Gaston Planté in 1859. Lead acid battery, which is based on the chemistry of lead, is well known among the available rechargeable batteries (Figure V.9).

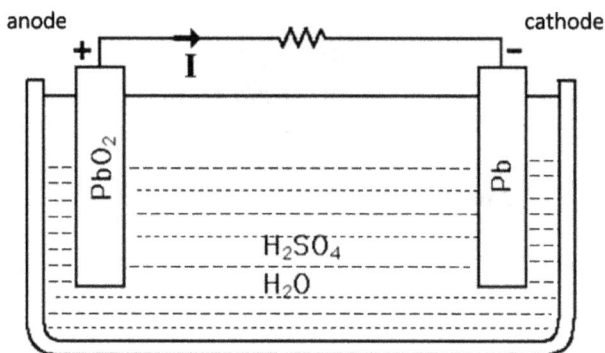

Figure V.9 A lead acid battery.

In this battery, lead serves as the anode and lead dioxide acts as the cathode, which are immersed into an electrolyte solution of sulfuric acid.

During charge: The sulfuric acid breaks up into positive hydrogen ions ($2H^+$) and sulphate negative ions (SO_4^-). If the two electrodes immersed in solution are connected to DC supply, the hydrogen ions being positively charged move towards negative terminal of the supply (cathode). The SO_4^- ions being negatively charged move towards positive terminal of the supply main (anode).

$$PbSO_4 + 2H_2O + 2H = PbSO_4 + 2H_2SO_4$$

During discharge: During this process, the electrodes are connected through a resistance, the cell discharges and electrons flow in a opposite direction. The hydrogen ions move to the anode and on reaching the anode receive one electron each from the anode and become hydrogen atoms.

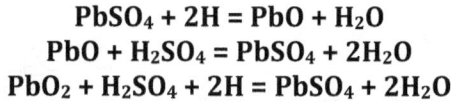

$$PbSO_4 + 2H = PbO + H_2O$$
$$PbO + H_2SO_4 = PbSO_4 + 2H_2O$$
$$PbO_2 + H_2SO_4 + 2H = PbSO_4 + 2H_2O$$

The lead acid batteries have following disadvantages:

* The requirement of inactive compounds such as grids, separators, cell container, etc., and high cost limits the utilization of lead acid battery.
* The actual value of specific energy (Wh/Kg) is only 25% of the theoretical one.
* To enhance the ionic conductivity in the charged and discharged states, an excess acid is necessary.
* The battery has a short life time.
* It requires high maintenance and has inadequate energy density.

Li-ion battery: Li-ion batteries are secondary batteries. The battery consists of an anode of lithium, dissolved as ions, into carbon. The cathode material is made up from lithium liberating compounds, typically the three electro-active oxide materials

* lithium cobalt oxide ($LiCoO_2$)
* lithium manganese oxide ($LiMn_2O_4$)
* lithium nickel oxide ($LiNiO_2$)

The lithium battery is constructed by coupling positive electrode with negative electrode in ion conducting electrolyte. The lithium ions move back and forth between the positive and negative electrodes through the electrolyte during the charge and discharge process. The positive electrode accommodates Li ions during the discharging of the battery and the negative active materials accommodate these ions during charging (Figure V.10).

During the charge and discharge processes, lithium ions are inserted or extracted from interstitial space between atomic layers within the active material of the battery. Simply, the Li-ion transfers between anode and cathode through lithium electrolyte. Since neither anode nor cathode materials essentially change, the operation is safer than that of a lithium metal battery.

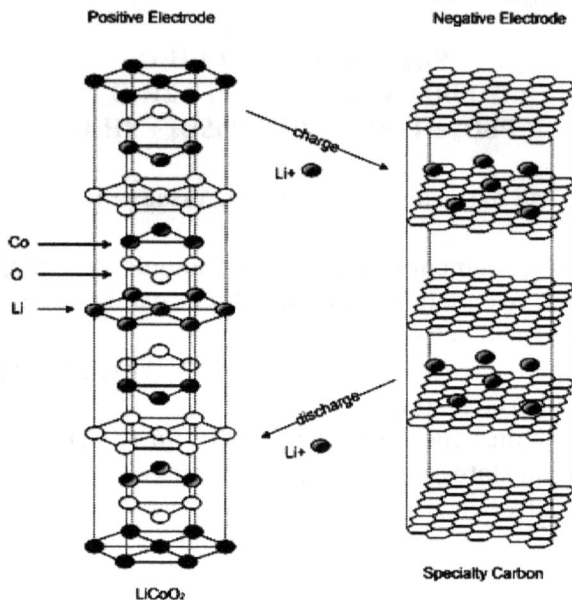

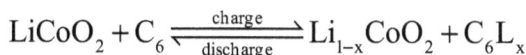

$$LiCoO_2 + C_6 \underset{discharge}{\overset{charge}{\rightleftharpoons}} Li_{1-x}CoO_2 + C_6L_x$$

Figure V.10 Operation of a Li-ion battery.

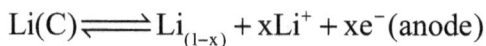

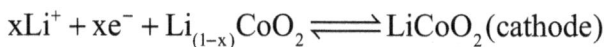

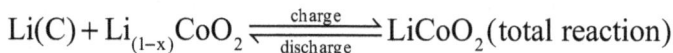

$$Li(C) \rightleftharpoons Li_{(1-x)} + xLi^+ + xe^- (anode)$$

$$xLi^+ + xe^- + Li_{(1-x)}CoO_2 \rightleftharpoons LiCoO_2 (cathode)$$

$$Li(C) + Li_{(1-x)}CoO_2 \underset{discharge}{\overset{charge}{\rightleftharpoons}} LiCoO_2 (total\ reaction)$$

The traditional batteries are based on galvanic oxidation-reduction reactions, but lithium-ion secondary battery follows an "intercalation" mechanism. The insertion of lithium ions into the crystalline lattice of the host electrode without changing its crystal structure takes place, hence, known as intercalation host. These electrodes should have two important characteristics such as these should have open crystal structure, which allow the insertion or extraction of lithium ions and the ability to accept compensating electrons (Figure V.11).

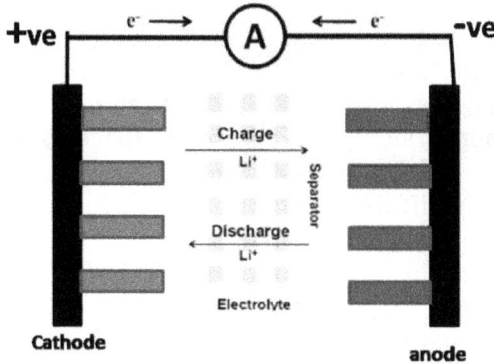

Figure V.11 Li ion battery.

The process is completely reversible. Thus, the lithium ions pass back and forth between the electrodes during charging   and   discharging. Due to this reason, the lithium-ion batteries are called 'rocking chair', 'swing' cells. A typical Li-ion battery can store 150 Watt-hours of electricity in 1 kg of battery as compared to 25 Watt-hours of electricity/1 kg in lead acid batteries. All rechargeable batteries suffer from self-discharge when stored or not in use. Normally, there will be a three to five percent of self-discharge in lithium-ion batteries for 30 days of storage.

The Li-ion batteries have following advantages:

- ❖ high energy density than other rechargeable batteries
- ❖ light weight
- ❖ produce high voltage as compared to other batteries
- ❖ improved safety, i.e., more resistance to overcharge
- ❖ no liquid electrolyte means these are immune from leaking
- ❖ fast charge and discharge rate

Some of the disadvantages of Li-ion batteries are:

- ❖ expensive
- ❖ not available in standard cell types

A few applications of Li-ion batteries include:

❖ The Li-ion batteries are used in cameras, calculators, etc.
❖ They are used in cardiac pacemakers and other implantable devices
❖ These are used in telecommunication equipment, instruments, portable radios and TVs, pagers, etc.
❖ The batteries are used to operate laptop computers and mobile phones, along with use in aerospace applications

V.10.2 Supercapacitors

Supercapacitors represent an important class of energy storage devices, particularly useful for short-acting high power batteries. Batteries suffer from a relatively slow power charge and discharge. To meet the demands of faster and higher-power energy storage systems for a number of applications, supercapacitors often replace conventional batteries.

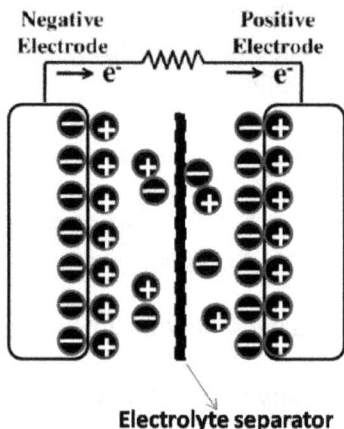

Figure V.12 A supercapacitor system.

Supercapacitors are formed by two polarizable electrodes, a separator and an electrolyte, as shown in Figure V.12. These are formed by the double layer capacitance of the ions of the electrolyte adsorbed on the charged electrode, resulting in a Helmholtz layer. Currently, the power density of supercapacitors is generally lower than that of batteries. Therefore, the development of supercapacitors is aimed towards improvement of power density and cost reduction. These high performance electrochemical devices are also known as ultracapacitors, pseudo-capacitors as well as double layer capacitors.

V.11 Fuel Cells

A fuel cell is a electrochemical device which uses a fuel (hydrogen or hydrogen-rich molecules) and oxygen to create electricity by an electrochemical reaction. It generates electricity inside a cell through reactions between a fuel and an oxidant, triggered in the presence of an electrolyte.

Fuel cells are different from conventional electrochemical cell batteries as these consume reactant from an external source, which must be replenished thermodynamically. By contrast, batteries store electrical energy chemically and, hence, represent a thermodynamically closed system.

V.11.1 H_2-O_2 Fuel cell

H_2-O_2 fuel cell is one of the simplest and most successful fuel cells (Figure V.13).

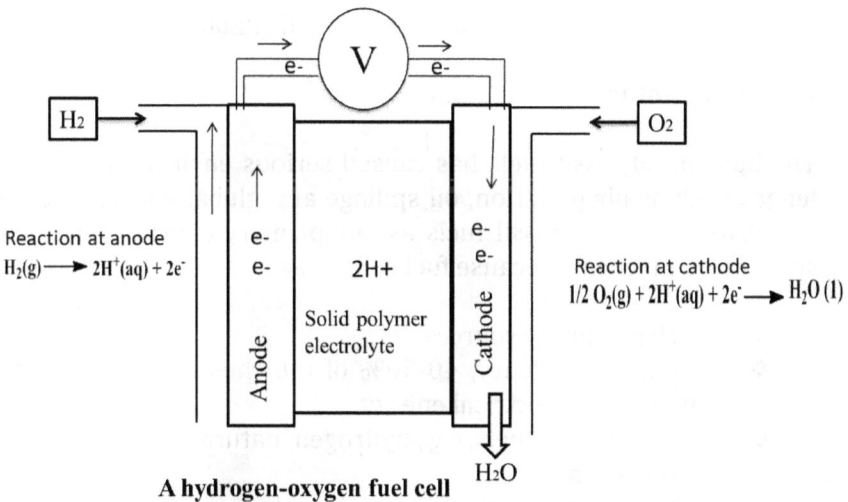

Reaction at anode
$H_2(g) \longrightarrow 2H^+(aq) + 2e^-$

Reaction at cathode
$1/2\,O_2(g) + 2H^+(aq) + 2e^- \longrightarrow H_2O\,(l)$

A hydrogen-oxygen fuel cell

Figure V.13 Hydrogen-oxygen fuel cell.

Assembly

It consists of two porous carbon electrodes as anode and cathode. These porous electrodes are made of carbon with a small amount of catalyst such as Pt, Pd and Ag. Two electrodes are connected through a voltmeter. Fuel H_2, oxidizer O_2 and liquid electrolyte or solid polymer electrolytes are employed for operation.

Hydrogen, or a hydrogen-rich fuel, is fed to the anode where a catalyst separates hydrogen's negatively charged electrons from positively charged ions (protons). At the cathode, oxygen combines with electrons and, in some cases, with species such as protons or water, resulting in water or hydroxide ions, respectively.

V.11.2 Types of Fuel Cells

There are five major types of fuel cells used in the market. All of these have the same basic design as mentioned above, however, different chemicals are used as electrolytes in these cells. These fuel cells are:

- ❖ Alkaline fuel cell (AFC)
- ❖ Phosphoric acid fuel cell (PAFC)
- ❖ Molten carbonate fuel cell (MCFC)
- ❖ Solid oxide fuel cell (SOFC)
- ❖ Proton exchange membrane fuel cell (PEMFC)

V.11.3 Advantages of Fuel Cells

The burning of fossil fuels has caused serious environmental challenges such as air pollution, oil spillage and global warming. Using fuel cells to replace fossil fuels as our primary energy source can solve these problems because fuel cells:

- ❖ are clean energy source
- ❖ have high efficiency, 40-50% of the chemical energy being converted to electrical energy
- ❖ use a variety of fuels, e.g., hydrogen, natural gas, methanol or hydrocarbons
- ❖ are reliable, maintainable and durable.

Practice Questions

UNIT I

WATER AND ITS TREATMENT

1. What is meant by hardness of water?
2. What is the role of EDTA?
3. What is the principle involved in the determination of total hardness of water by EDTA method?
4. How to estimate the hardness of water?
5. Distinguish between:
   a) temporary and permanent hardness
   b) sludge and scale
   c) softening and demineralization
6. Explain the treatment of boiler feed water.
7. What are the functions of lime and soda in hot lime-soda process? Give equations.
8. Why does hard water consume more soap?
9. Which problems can arise in boilers? Why are they caused? What are the methods of their elimination?
10. What is boiler feed water? What are its requirements?
11. What is a zeolite? Give an example.
12. Define ion-exchange process?
13. Briefly explain the internal treatment of boiler feed water.
14. Compare internal and external treatment of boiler feed water.
15. Explain the desalination of brackish water.
16. What is meant by reverse osmosis?
17. What is desalination? Name the different methods of desalination and describe any one.
18. Write notes on: (i) caustic embrittlement, (ii) reverse osmosis, (iii) priming and foaming, (iv) hot lime-soda process and (v) boiler corrosion.

UNIT II

SURFACE CHEMISTRY AND CATALYSIS

1. Define adsorption.
2. What is the difference between adsorption and absorption?
3. Define adsorption isotherms.
4. Why adsorption is exothermic in nature?
5. Explain the Freundlich's adsorption isotherm.
6. Derive the Langmuir's adsorption isotherm.
7. What is the effect of temperature on adsorption?
8. Physisorption is reversible while chemisorption is irreversible, why?
9. Define contact theory?
10. Write about the applications of adsorption for pollution abatement.
11. Derive rate expression for the unimolecular adsorption reactions on surfaces.
12. What is meant by catalysis?
13. Explain(i) autocatalysis and (ii) catalytic poisoning.
14. Explain acid-base catalysis.
15. Define enzyme catalysis.
16. Derive the Michaelis-Menten equation.

UNIT III

ALLOYS AND PHASE RULE

1. What is an alloy? Give examples of ferrous and non-ferrous alloys.
2. Write about composition and uses of bronze.
3. State the significance of increasing carbon content in steel.
4. Explain the synthesis and properties of nichrome and stainless steel.
5. What are the functions and effects of alloying elements?
6. What is the significance of alloying elements?
7. State phase rule and explain the terms involved in it.
8. Explain the heat treatment processes: a) annealing, b) tempering and c) hardening.
9. Explain the heat treatment processes: (i) nitriding, (ii) normalizing and (iii) carburizing.
10. Define reduced phase rule.
11. Briefly explain the thermal analysis and cooling curves. What are the uses of cooling curves?
12. What is meant by phase rule?
13. Define phase, components and degrees of freedom.
14. Calculate the number of components

$$CuSO_4.5H_2O(s) \rightleftharpoons CuSO_4.H_2O(s) + 4H_2O(v)$$

$$PCL_5(s) \rightleftharpoons PCl_3(v) + Cl_2(v)$$

15. Calculate the number of phases in the following:

$Sulphur(monoclinic) \rightleftharpoons Sulphur(rhombic) \rightleftharpoons Sulphur(liquid)$

$Water + Alcohol \rightleftharpoons Vapour$

16. Discuss the phase diagram of lead-silver system.
17. Explain Pattinson's process.
18. Discuss one component system of water with a neat phase diagram.
19. What is condensed phase rule? State its significance.
20. What is meant by triple point? State its characteristics.
21. What is an eutectic point?
22. Discuss the effect of Ni, Cr and Mn in the alloying of steel.

UNIT IV

FUELS AND COMBUSTION

1. Define fuel.
2. What are the advantages and disadvantages of solid fuels?
3. How are the gaseous fuels superior to other fuels?
4. What are the requirements of a good fuel?
5. Explain the gross and net calorific value. How are they related?
6. What is metallurgical coke?
7. Explain the manufacture of metallurgical coke by Otto-Hoffmann method and the recovery of various byproducts.
8. How is synthetic petrol obtained by Bergius method?
9. What are GCV and LCV of a fuel?
10. How will you increase the rate of combustion?
11. What is carbonization?
12. How is coke superior to coal?
13. What is meant by refining of petroleum?
14. What is cracking?
15. What is knocking? How is it rectified?
16. What is octane number? How is it improved?
17. Define cetane number.
18. Distinguish between petrol and diesel.
19. What is LPG? Mention its chemical composition and calorific value.
20. What is meant by proximate and ultimate analysis?
21. State the composition and use of producer gas.
22. Mention the significance of flue gas analysis.
23. Describe the proximate and ultimate analysis of coal and their significance.
24. What is cracking? How is it useful of the preparation of synthetic petrol?
25. What are the characteristics of good metallurgical coke?
26. What is crude oil? What are the various fractions obtained by the fractional distillation of crude oil? Mention the composition and uses.
27. How is flue gas analysis carried out? Explain with neat diagram.

UNIT V

ENERGY SOURCES AND STORAGE DEVICES

1. What is meant by nuclear fission?
2. What is meant by nuclear fusion?
3. Explain differences between nuclear fission and fusion.
4. Illustrate nuclear chain reaction.
5. Explain nuclear chain reaction.
6. Define nuclear energy.
7. Define batteries and provide examples.
8. Briefly explain lithium-ion-battery.
9. Explain the primary and secondary batteries with examples.
10. What is meant by supercapacitors? State their importance.
11. What are the uses of fuel cells?
12. Discuss the characteristics of wind energy.
13. List various advantages of wind energy.
14. Draw the schematic of lead-acid battery, provide its cell reactions.
15. Write a short note on solar cells.
16. What are the advantages of solar energy conversion?
17. Explain the charge and discharge reaction of Li ion battery.
18. Discuss the importance of breeder reactor.

Index

www.ingramcontent.com/pod-product-compliance
Lightning Source LLC
Chambersburg PA
CBHW071704210326
41597CB00017B/2331